# Millennium Stocks

### RICHARD C. DORF, Ph.D.

**University of California, Davis**

$S^{t}_{L}$

**St. Lucie Press**
Boca Raton   London
New York   Washington, D.C.

**Library of Congress Cataloging-in-Publication Data**

Dorf, Richard C.
    Millennium stocks / by Richard C. Dorf
        p.  cm.
    Includes bibliographical references and index.
    ISBN 1-57444-250-3 (alk. paper)
    1. Stocks. 2. Investments. I. Title.
HG4661 .D67 1999
332.63'2221—dc21
                                                         99-045048
                                                            CIP

# Preface

All investors seek a superior return on their investments in common stocks by selecting the stocks of companies that have superior profitability and return on investment. Stock prices are influenced by the expectation of earnings, the consistency of their earnings, and the expected growth of these earnings over the investment period. Furthermore, the value of a company's shares can be calculated as the cash that investors can expect to get back over the investment period, adjusted for the risk of the investment.

The successful firm returns value to its shareholders when the return on the invested capital exceeds the cost of the invested capital. Thus, the investor should seek companies that evidence the ability to generate superior profitability and return on capital and the demonstrated ability to sustain this growth and profitability. We call these companies "powerful companies" since they possess sustainable power to generate value.

Powerful companies possess the ability to generate profitable cash flow even when the economy experiences a downturn as in 1990–91. The real value of a powerful company cannot be determined solely by traditional accounting measures. The worth of Microsoft or General Electric lies not only in its physical assets or inventories but in its intellectual capital. The source of wealth is information, know-how, knowledge and intellectual assets such as patents, secrets and new processes of operation.

The purpose of this book is to provide a road map towards the identification of valuable companies that are fairly priced and are predicted to grow in wealth due to their intellectual capital. In the book, we develop a formula for power and value that will assist the investor in finding great companies with a solid future. The reader then builds a portfolio of these stocks which we call Millennium Stocks. This portfolio is built with knowledge of the risk and holding period for the portfolio. We identify and list for the reader 100 Millennium stocks based on their expected return on capital and growth of cash flow, while accounting for the risk.

# Acknowledgements

I sincerely acknowledge the wise comments and insights of my colleagues at the University of California, Davis: Brad Barber, Paul Griffin and George Bittlingmayer. The efforts of my editor, Drew Gierman, are sincerely appreciated.

# Table of Contents

# About the Author

Richard C. Dorf is professor of management in the Graduate School of Management at the University of California, Davis. Professor Dorf is the author of the *New Mutual Fund Advisor, The Mutual Fund Portfolio Planner, the Handbook of Technology Management,* and *The Manufacturing and Automation Handbook.*

Professor Dorf has extensive experience with education, industry, and finance. He has served as adviser to numerous firms, financial institutions and governmental agencies. He has served as a visiting professor at the University of Edinburgh, Scotland, the Massachusetts Institute of Technology, Stanford University, and the University of California, Berkeley.

# Introduction

## The Twelve-Step Process of Creating Long-Term Wealth

### New Wealth

By any measure, the effect of change in organizations and the marketplace has grown significantly. Technology is now central to change and new cultural modes lead to dynamic changes in our lives. As these changes spread throughout the world, a reordering of the economy occurs. The dynamics of the rearranged economy leads to new wealth opportunities. The new economy is global and favors innovation in the use of information and communication. As the world of intangibles, media, software, and services grows in importance, new economic opportunities evidence themselves in the stock market. It is the key premise of this book that new wealth will flow from the innovations taking place in health care, communication, and technology. Thus, our search for long-term generation of wealth is fostered by emphasizing investment in the most attractive segments of the economy while maintaining a long-term, diversified portfolio of equity holdings primarily in U.S. companies.

The value of U.S. households' portfolios of equity stocks and mutual funds rose 20% to $10.8 trillion in 1998 and now represents 25% of household assets.

Stocks (equity securities) of U.S. companies represent a part of the strong U.S. economy. U.S. companies have the highest returns on invested capital. This has resulted from sound management practices and productive methods of operating companies as well as excellent opportunities for innovative products and services. Thus, in this book we will seek to identify strong, well-managed U.S. companies with a solid franchise in their sector of the economy.

### Twelve Steps

The 12-step process shown in Exhibit 1 should enable the reader to build and maintain a reasonably diversified portfolio of U.S. equities that may in the long-run match or exceed the return of the S&P 500 stock index with low tax consequences. This twelve step process is explained in Chapters 1 through 15. Using the companies identified in Chapters 16 and 17, the reader can build a portfolio and a strategy for

**Exhibit 1    The Twelve-Step Process of Creating Long-Term Wealth**

1. The Economy.
   - Examine the emerging economy in the U.S.
   - Identify models and scenarios of the economy.
   - Identify the opportunities for the future.

2. Intellectual Capital.
   - Identify the role of intellectual capital.
   - Determine models for intellectual capital companies.
   - Determine profit models for growth companies.

3. Risk.
   - Risk versus reward.
   - Diversification.
   - Timing.

4. An Investment Plan.
   - Size of companies.
   - Trading.
   - Asset allocation.

5. Sectors of the Economy.
   - Identify the characteristics of the sectors.
   - Identify and describe the best opportunity sectors.

6. Select Candidate Companies in Each Sector.
   - Select high growth companies.
   - Select medium growth companies.
   - Select growth and income companies.

**Exhibit 1 The Twelve-Step Process of Creating Long-Term Wealth (continued)**

7. Select Power Stocks.
   ■ Use power stock criteria.

8. Select Contrarian and Technology Stocks.
   ■ Select contrarian stocks using Contra criteria.
   ■ Develop a valuation technique for potentially high reward technology stocks.

9. Build a Long-Term Portfolio Strategy.
   ■ Identify a risk-reward balance.
   ■ Develop an asset allocation balance.
   ■ Build a portfolio of growth, income, and Contra stocks.
   ■ Consider high reward stocks.

10. Prepare for Bear Markets.
    ■ Develop a strategy.

11. Build a Buying and Selling Tactic.
    ■ Determine a buying tactic.
    ■ Determine a selling tactic.

12. Review and Revise the Millennium Portfolio.
    ■ Let winners grow and sell laggards.
    ■ Add new emerging growth companies.

creating wealth. First, the reader will develop an understanding of the U.S. economy and the opportunities for the future (Chapter 1) and then identify the role of intellectual capital in our economy so that companies with significant intellectual capital can be identified in Chapter 2 (Step 2).

In Chapter 3, we define risk and portfolio diversification and discuss the balance between risk and reward. In Chapter 4 we build an investment plan for superior returns and moderated risk.

Then, in Step 5, the reader examines the sectors of the economy that will afford the best investment opportunities for the future. The next step (Chapters 5 and 6) is to develop a technique for valuing candidates in each attractive sector using high growth and medium growth companies. The power stock criteria are described in Chapter 7 (Step 7) and illustrated for several outstanding performers.

Step 8 focuses on selecting turnaround candidates for a contrarian segment of the total portfolio and also selecting high growth technology stocks. Step 9 uses a risk-reward measure so that the reader can build a portfolio of growth, income, and Contra stocks. Step 10 helps the reader prepare a strategy for a bear market. Step 11 helps the reader develop a buying tactic and a selling tactic for her portfolio.

Chapter 12 helps you prepare for global investing. Chapters 13 and 14 respectively help you select technology stocks and value stocks. Chapter 15 prepares you for market declines. Chapter 16 identifies 100 Millennium stocks using the power criteria. Finally, Chapter 17 provides data on the 100 Millennium stocks that will enable you to further examine their potential.

Step 12 periodically reviews the Millennium portfolio and lets the stock winners grow while selling laggards and adding new emerging growth companies.

## *Stocks—Not Mutual Funds*

In this book we utilize stocks of leading companies to build a portfolio that can provide a good return over the long run while maintaining a volatility of annual return that the reader can accommodate. We use a stock portfolio instead of mutual funds because the advantages of a stock portfolio are compelling. If you plan to have a portfolio larger than $100,000 you will be better served using your own stock portfolio as long as you have the time and inclination to learn about the companies in your portfolio.

**Exhibit 2   The Benefits and Costs of Mutual Funds versus a Stock Portfolio**

<table>
<tr><th>Advantages</th><th>Disadvantages</th></tr>
<tr><td colspan="2" align="center">**Mutual Funds**</td></tr>
<tr>
<td>
■ Professional manager<br>
■ Readable quarterly reports<br>
■ Ready exchange among funds<br>
■ Diversified fund
</td>
<td>
■ Annual fee and costs of 1.5% or more<br>
■ 80% of funds do not have returns equal to the S&P 500<br>
■ Turnover rates of 80% or greater<br>
■ Annual tax consequences<br>
■ Often unclear what the fund actually uses for a strategy or style<br>
■ Trading costs may exceed 1% annually
</td>
</tr>
<tr><td colspan="2" align="center">**Stock Portfolio**</td></tr>
<tr>
<td>
■ Keep close to the market<br>
■ Control of tax consequences<br>
■ Increased confidence during a market downturn<br>
■ Significant knowledge of the companies in the portfolio
</td>
<td>
■ Time required<br>
■ Less diversification<br>
■ Research necessary for stock choice<br>
■ Maintain records of individual stocks
</td>
</tr>
</table>

# 1 The Economy

## This chapter covers

- The Large Corporation
- The Technological Revolution
- The Entrepreneurial Momentum
- Inflation and Deflation
- Interest Rates
- The Business Cycle
- Productivity and Growth
- Models of the Economy

## The Large Corporation

The characteristic institution of today's market economies is the large business corporation. It is the vehicle for the technologies of production and distribution and the organizational arrangement that provides the engine of productivity. Globalization of markets has resulted in the need for large companies competing internationally. The global market has provided opportunities for efficiencies, scale, and profit to any company big enough to take advantage of them. Coca-Cola and Gillette are excellent examples of large global companies. In the U.S., deregulation has led to a substantial increase in competition in the transportation and telecommunication markets and in some parts of the markets of gas and electric public utilities. The successful development of the cellular telephone industry is due to large, well-capitalized companies like Motorola and Lucent. Large, innovative companies use their resources and workers' creative skills to adapt to change. Merger and acquisition activity is active within nations and between nations as seen in the merger of Daimler-Benz with Chrysler to form Daimler-Chrysler. There is a worldwide trend towards large companies and consolidation of companies. In the new world of finance and industry, size counts. Big companies enjoy economies of scale and name recognition, and they can be safer because of their deep resources and the diversity of their markets. Big companies are better able to marshal the capital and research funds, build a global brand and convince distributors and retailers to carry their product. We live in an economy where innovation matters most and larger companies must emphasize creativity and imagination. In this book we choose to primarily focus on the large company with a market capitalization of $1 billion or greater.

# The Technological Revolution

The U.S. economy is led by technological advances that enable increased economic growth. From the internet to biotechnology, many new innovations result in increasing productivity and living standards. The information revolution will continue to boost productivity in information-dependent industries including finance, media and wholesale and retail trade. It is important to choose the technologies that actually boost productivity. Examples of commercial failures of innovation are nuclear energy and much of biotechnology. Biotechnology may yet prove fruitful as new commercial applications emerge. Many of the high-tech industries may grow revenues at 15% per year while the prices of their goods fall yearly. Thus, in these industries, we often experience economic growth with accompanying disinflation. Corporations are investing in computers, communications, networks and other labor-saving technologies in order to boost productivity.

Since the 1850s the U.S. has gone from one new technology to another: railroad, telegraph, electric power, automobile, telephone, radio, transistor, internet.

The service sector of the U.S. economy accounts for 75% of output and employment. Services including insurance, finance, entertainment, and education are favorably improved by the use of productivity-enhancing technologies. The dominant technologies of the next decade will be microelectronics, telecommunications and biotechnology. The elements of the U.S. economy are summarized in Exhibit 1.1.

# The Entrepreneurial Momentum

Entrepreneurialism and individual initiative in the U.S. are so widely accepted that approximately 2 million businesses were started in the 1990s. Small companies and new venture units of larger companies create new business through innovation with their capacity to adapt and attract new creative people. From the small business startups have emerged the companies that have provided the energy and jobs in our economy. The new bottom-up business environment enables innovation to emerge from the small firm as well as the new venture started by a large company.

The **entrepreneur** is a person who seeks to create wealth and lasting value by starting and developing a new business venture. The entrepreneur can start a new independent business or start, with the support of his employer, a new venture as a corporate spinoff or internal corporate venture. An excellent example of a modern entrepreneur is Michael Dell.

### Exhibit 1.1   The Elements of the Economy

- Large companies, market capitalization greater than $1 billion.
- The technological revolution based on communications and computing.
- Steadily increasing productivity.
- Globalization of business.
- The service sector.
- The entrepreneur and small developing companies
- Inflation and Deflation
- Interest Rates
- The Business Cycle

The entrepreneur excels in a period of dislocation as an economy evolves from one business wave to another. The agricultural wave lasted until the economy shifted towards the industrial wave during the transition period 1800–1850. The industrial economy lasted through the period 1850–1960. The next transition period was from 1960–2000 as the information wave emerged.

Entrepreneurism is something Americans do best. New business incorporations rose to 900,000 in 1998. The most obvious entrepreneurs are the technologists, but others have started restaurants such as Starbucks and bookstores such as Amazon.com. The U.S. Labor Department states that new companies create 60% of all the new jobs.

## Inflation and Deflation

**Inflation** is a sustained increase in the prices of goods and services, usually expressed as an annual percentage increase. Conversely, **deflation** is a sustained decrease in prices of goods and services. **Disinflation** is a general declining trend of the annual rate of inflation. During the 1970s, the U.S. rate of inflation increased. Since the mid-1980s we have experienced disinflation as shown in Table 1.1. The forces for inflation are low economic growth and excessive money creation. The forces for disinflation are falling prices, improved productivity, lower import costs and overcapacity in parts of the world. General price inflation retards economic growth since it can lead to a misallocation of resources. In the period 1990–1999, the U.S. economy experienced disinflation with deflationary forces evident due to excess capacity in several industries. Deflationary forces may harm old industries like autos and agriculture whose products do not change much over time and act like commodities. During a period of disinflation, investors reduce their inflation expectations, often leading to lower demanded interest rates. During periods of disinflation, companies emphasize productivity growth since the power to increase prices is low or nonexistent. In the period of the postwar (1952–1965) growth, U.S. inflation rose at a mere 1.3% annual rate. Inflation could remain tame for many years to come.

There is the possibility that the U.S. may actually experience a period of deflation (falling prices of goods and services). Some deflationary forces include: deregulation, technology, decreasing government spending, increasing global competition, increasing strength of the dollar, a contraction in the money supply, overcapacity, and a switch from borrowing and spending to saving. If many of these forces converged, deflation could occur for one or more years. Winners and losers in a deflationary economy are contrasted with the winners and losers in an inflationary economy in Table 1.2. The best we can hope for is a very low, stable inflation/deflation balance within a range of ±1%.

**Table 1.1   Consumer Price Index, Year to Year % Change**

| Year | '74 | '77 | '80 | '81 | '85 | '90 | '94 | '98 | '99 (Est.) |
|------|-----|-----|-----|-----|-----|-----|-----|-----|------------|
| CPI | 12.0% | 6.0% | 14.0% | 10.0% | 4.0% | 4.3% | 2.5% | 1.5 % | 1.8% |

**Table 1.2 Winners and Losers in a Deflationary Economy and an Inflationary Economy**

|  | Inflationary | Deflationary |
|---|---|---|
| Winners | ■ New technologies<br>■ Real estate<br>■ Commodities producers<br>■ Natural resources | ■ New technologies<br>■ Bonds (long-term)<br>■ Low-cost producers<br>■ Services to upscale customers<br>■ Productivity enhancers |
| Losers | ■ Bonds (long-term)<br>■ Savers<br>■ Holders of cash | ■ Real estate<br>■ Capital intensive firms<br>■ Old technologies<br>■ Autos, housing<br>■ Commodities |

## Interest Rates

Interest rates have declined as the rate of inflation has declined in the period 1980 to the present (see Table 1.3). Low rates are a boon to nearly everyone. Cheap money aids the homeowner in refinancing and the business owner in expansion and working capital loans. Usually interest rates rise through the course of an economic expansion as the economy starts to overheat and labor pushes for wage increases. With a balanced federal budget and disinflationary pressures, interest rates may remain tame for several years to come. Often the interest rate of the 30-year Treasury Bond is used as the discount rate to determine the value of a stock. The 30-year Treasury Bond rate for several periods is given in Table 1.3.

## The Business Cycle

The periodic ups and downs of the economy around its long-term trend is called the **business cycle.** Recurring patterns of expansion and contraction occur in particular industries. The business cycle of an industry will have an effect on the overall economy, but an industry and the economy do not always move synchronously. A **recession** is a period of decline in total output, income, employment and trade, and is marked by widespread contraction in many sectors of the economy. Often, recessions or contractions occur in times of cold war, regulation, excessive taxation, or unstable monetary policy. Factors that moderate business cycles are listed in Table 1.4. The recessions of 1973–74 and 1991 were caused, in part, by a sharp increase in the price of oil as well as strong challenges to world peace. Capitalism is driven by cycles of creation and destruction. The system needs periodic shakeouts and corrections to eliminate inefficiencies. Thus, cycles will always exist and the investor should build a method to ride out the downturns.

Expansions tend to succumb to one of three forces. The most common in recent decades is an overheating economy with rising inflation. This causes the Federal

**Table 1.3 The 30-Year Treasury Bond Yield**

| Year | '60 | '74 | '80 | '85 | '90 | '98 | '99 (Est.) |
|---|---|---|---|---|---|---|---|
| **Interest Rate** | 2.9% | 7.0% | 16% | 9.0% | 8.3% | 5.25% | 5.90% |

**Table 1.4   Moderating Factors on the Business Cycle**

- Deregulation
- Peace
- Tax moderation
- Less inventory buildup
- Stable monetary policy
- Broader outsourcing to a variety of vendors
- Spread of capitalism
- Rising share of the economy from services

Reserve to raise interest rates, which chokes off growth. A second common force is an external shock, such as a sharp rise in oil prices, as in 1973–74. The third force is a financial crash when a speculative financial bubble bursts as occurred in Japan in 1988–1992.

## Productivity and Growth

The forces facilitating long-term economic growth in the U.S. are: transparent financial markets; well-capitalized, well-regulated banks; free trade; educated workers; a reliable but not inflexible legal system; taxes and welfare benefits low enough to avoid disincentives to work. Today it is the lean and flexible American economic model that is deemed triumphant, as the U.S. enjoys rapid growth, low unemployment and low inflation.

**Productivity** can be defined as the ratio of output divided by input. Growth of productivity measured in Gross Domestic Product (GDP) output per hour of work input is one of the most important measures of the economy. A lasting rise in productivity can mean bigger wage or salary increases without inflation and a bigger GDP.

After World War II, nonfarm business productivity grew 2.8% a year on average. The pace slowed around 1973, and for the next 20 years, productivity growth averaged only 1% a year. Since 1994, however, the pace has quickened and productivity has improved by 2% a year since 1995.

Another measure of economic performance is growth in GDP per capita which has averaged 1.6% in America over the past decade. Economic studies suggest that the best recipe for growth is a long list, but with no single magic ingredient. The list includes high saving, support of entrepreneurs, innovation, low taxes, openness to trade, good education, the rule of law, and sound monetary and fiscal policies.

Many economists assert that the economic value of quality, speed, customer service and new products is not captured by government statistics such as GDP. This problem is complicated by the challenge of defining and measuring output in the service sector, which includes finance, banking, health care, and education.

## Models of the Economy

An investor builds a mental model of the economy based on appropriate information sources and uses this model to continually examine the economy and adjust expectations of returns as well as review investment opportunities. Based on the material considered in this chapter we examine the model shown in Exhibit 1.2. The investor

**Exhibit 1.2    Model of the Economy and the Stock Market**

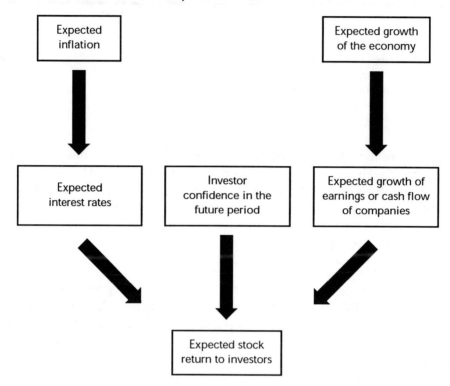

estimates the expected inflation and growth of the economy for the period under consideration (typically three to five years). It is also best to estimate the trends in inflation and growth as well as the levels. Most economic events are unpredictable and thus we recognize our expectations are fallible forecasts. Nevertheless, with a measure of confidence in a stable, moderated market condition, the investor records these expectations. This will lead to a model which can revisited periodically. One possible method is use the chart shown in Table 1.5 which is completed for the year 1999. With this model of the economy the investor then turns to other factors to determine the best candidates for stock purchase. In the next chapter we examine the role of intellectual capital in the success of modern corporations and determine several quantitative measures of attractive companies.

**Table 1.5    Expected Stock Return for 1999 for the S&P 500**

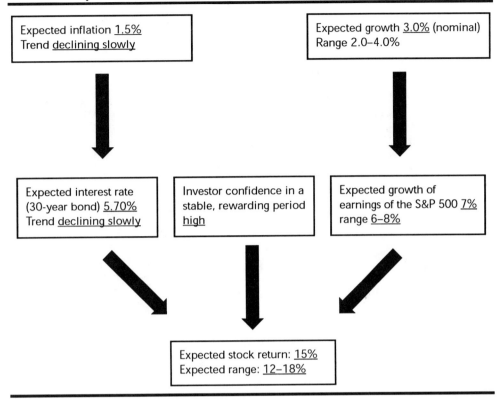

# 2 Intellectual Capital and Wealth Creation

## This chapter covers

- The Definition and Role of Intellectual Capital
- Increasing Returns
- Measures of Intellectual Capital
- Profit Models and Value Migration

## The Definition and Role of Intellectual Capital

The economy of the new millennium will be defined by the growing value of knowledge as an input and output and the rising importance of intellectual capital relative to real estate, physical plant and equipment, and financial capital. The ability to manage knowledge assets is critical to wealth building companies. Wealth is not easily developed with commodities such as forests, paper, and grains although new materials and value-added products made from commodities will create wealth. Wealth flows from significant value-added and value-added flows from knowledge and intellectual capital. The old foundations of success are eroding. In the past the method for success was to control natural resources. Now the answer is to control knowledge.

Many, if not all, competitors in a given industry can access the new machines necessary to upgrade performance. It is not, however, the machines that differentiate the best firms, but rather how they are employed. Knowledge is the most important factor of production and managing intellectual assets has become the focus of business. The inputs to the processes of a business are raw materials, energy, capital, and labor, but the most import input is how we deploy these assets using intellectual assets. The productive business provides reformed and converted resources using intellect and knowledge to provide a product or service. (See Exhibit 2.1.)

The current business transition is towards more value-added processes. **Value added** is the difference in value between the inputs and the output, and is dependent on the intellectual capital embedded in the process. When the research and development (R&D) investment of a firm begins to surpass the capital investment, the corporation is shifting to an intellectual company.

**Exhibit 2.1   Intellectual Capital and the Business Process**

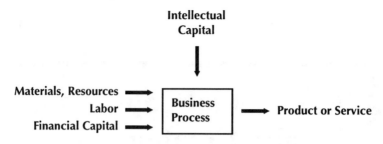

The **intellectual capital** of a firm includes the talents of its people, the efficacy of its management systems, the effectiveness of its customer relations, and the techno-logical knowledge employed and shared among its people and processes. Intellectual capital is knowledge that has been formalized, captured and used to produce a process that provides a significant value-added product or service. Intellectual capital is use-ful knowledge that has been recorded, explained, and disseminated and is accessible within the firm [Stewart, 1997]. Of course, in order to retain this capacity to add value, firms make efforts to keep the existing knowledge within their walls and work to build new knowledge so as to remain in the lead.

It is the job of a business to build corporate assets of which the critical asset is knowledge. The use of patents, databases, design methods, software codes, brands, hardware designs, manufacturing methods, machines, and tools are all examples of knowledge assets.

**Brand equity** is one part of intellectual capital that can have lasting power. Brand equity is the collection of all assets linked to a brand, its name and symbol that add to the value provided by a product or a service. Brand equity can have a direct influence on the market value of a company. Examples of companies with high brand equity are McDonald's and Coca-Cola.

Companies are said to be bundles of capabilities. Companies compete by doing things consistently better than the competition. This can be achieved with the right set of people, knowledge within the firm, and the coherence of processes binding all the people together with the knowledge of a mission represented by a clear vision. The company is a collection of capabilities based on precise knowledge. Like a person, a company is what it knows and what it is able to do with that knowledge.

In a dynamic environment, corporations that move with change are far sighted and focus on regenerating core competencies. A **competence** is a bundle of skills and tech-nology and represents the sum of learning by individuals and organizations [Hamel, 1994]. Hamel and Prahalad state that the ratio of market value to asset value is 2:1, 4:1, even 10:1, and that the difference between asset value and book value is not goodwill, it is the core competence—people-embodied skills.

The great opportunities for wealth creation occur where great disequilibriums or imbalances occur in the economy due to new circumstances. A new technology or method opens up opportunities to jump to new products or processes with enhanced capabilities and productivity. Examples are the introduction of electric power, the radio, and the internet. With electricity came lighting, electric motors, street railways, the telephone, and the radio. The introduction of the diesel locomotive is a 20th-century example of a large opportunity. The diesel locomotive is more powerful and efficient than the steam locomotive. The transition occurred rapidly in the period 1925–1945. Often the producers of the old technology cannot move to the new technology and ultimately lose the whole market. When breakthrough technologies come along, new firms often destroy the old firms.

The direct relationship between a company's book value and its equity price has declined over the past decades. Knowledge assets are not included in book value. Thus, we may loosely attribute the difference between book value and price as equal to intellectual capital.

The source of intellectual capital is twofold: human capital and organizational capital. Human capital is the combined knowledge, skill and ability of the company's employees. Organizational capital is the hardware, software, databases, methods, patents, and every element of the organization that supports the human capital [Edvinsson, 1997]. Therefore,

$$\begin{array}{c} \textbf{Intellectual} \\ \textbf{Capital} \end{array} = \begin{array}{c} \textbf{Human} \\ \textbf{Capital} \end{array} + \begin{array}{c} \textbf{Organizational} \\ \textbf{Capital} \end{array}$$

It is the role of the corporation to structure and manage the organizational capital and obtain and develop the best human capital in order to secure outstanding intellectual capital which leads to a value-added product or service and thus ultimately to wealth creation.

Major industries in our modern economy are built on a vast, growing network of systematized knowledge. New service, net technology, vast databases, computer simulations, and mathematical analysis come together in an infrastructure of knowledge.

## Increasing Returns

Many firms work within a decreasing return structure where marginal costs increase as increasing resources are added, therefore exhausting scale advantages (diminishing returns). In other words, the more a firm produces, the less profit per unit it achieves. Increasing returns for certain firms occur when marginal costs decrease as resources are added, thus enhancing returns. Increasing returns are the tendency for the firms which are ahead to get further ahead. The phenomena of increasing returns usually occurs in knowledge-based industries where intellectual capital is paramount. Positive feedback mechanisms support the winners and disadvantage the losers because the more a winner produces, the more profit per unit. Increasing returns are driven by the five factors shown in Table 2.1 [Teece, 1998]. The first factor is the development of a standard and a network built on that standard, such as Microsoft Windows. The second factor is the race to gain a lock-in of the customer so that the customer will be retained due to high switching costs. Third, large up-front costs of development will deter competitors if the first firm has locked up a large share of the market. This is very powerful, if the fixed cost of development is high and the variable unit cost is low or near zero. Finally, the fifth factor is getting the timing right so the market grows rapidly as the product is introduced widely.

The combination of these five forces is powerful.

**Table 2.1    Factors Supporting Increasing Returns**

1.  Ownership of standards and networks.
2.  Customer lock-in.
3.  Large up-front development costs and low unit costs to make.
4.  Organizational learning of tacit knowledge.
5.  Good timing.

Consider the internet provider America OnLine (AOL). It has become a standard with a large, rapidly growing network of over 15 million individual subscribers. The lock-in of customers is large since the switching costs may be high for most users. The development costs of this network exceed $100 million while the cost of adding a new subscriber is less than one ($1) dollar. Organizational tacit knowledge is the method and process of running this network and it is very difficult for a competitor to replicate. Finally, AOL has grown with the demand for the internet and e-mail and exhibits very good timing. Thus, AOL should experience increasing returns as it grows in number of users and thus can reap higher prices for its advertisements and services.

Increasing returns are best achieved by firms that meet all of the five factors listed in Table 2.1. In order to access a market with increasing returns the firm must develop its ability to create, transfer, assemble, integrate, and exploit intellectual capital. In order to capitalize on increasing returns companies often give away the first model of their product and provide the follow-up product at essentially zero cost to the producer and a reasonable price to the purchaser. Thus, AOL provided a disk and one month free to get customers on their service. After a month, the user pays a monthly fee.

## Measures of Intellectual Capital

Since intellectual capital is critical to a modern corporation, we need to identify companies with significant intellectual capital. Because intellectual capital is not accounted for on a company's balance sheet, we use indirect measurements of intellectual capital. In Table 2.3 we list several categories of intellectual capital. We only show measures that can be readily obtained from a report or common source of stock data. The four measures of intellectual capital shown in Table 2.2 are imprecise but useful. Each industry has one or two measures that are appropriate for the business conditions and reporting practices within that industry.

The four intellectual capital measures are shown in Table 2.2 for six companies for

**Table 2.2    Four Measures of Intellectual Capital**

1.  Sales per employee
2.  Cash flow per employee
3.  Sales divided by net property, plant, and equipment
4.  R&D divided by sales

**Table 2.3 Intellectual Capital Measures for Selected Stocks for 1998**

| Company | Sales Employee | Cash Flow Employee | Sales Net Property | R&D Sales | Industry |
|---|---|---|---|---|---|
| Cisco | $564,000 | $146,600 | 14.2 | 12.1% | Computer Hardware |
| Microsoft | $535,000 | $215,000 | 9.9 | 17.2% | Computer Software |
| Sara Lee | $144,000 | $12,400 | 9.6 | N.A. | Food Processing |
| Clorox | $498,000 | $79,200 | 4.6 | 2.0% | Household Goods |
| Analog Devices | $171,000 | $37,500 | 1.7 | 17.8% | Semiconductors |
| Deere | $374,000 | $35,300 | 8.1 | 3.2% | Farm Equipment |

1998. Each company shown is in a different industry. The sales per employee is a measure of the human capital of each firm. Cisco stands out as the highest on this measure.

Cash flow per employee is also a measure of human capital and serves as a good measure of profitability. Sales divided by net property, plant and equipment is a measure of sales generated relative to investment in physical assets. Cisco and Microsoft have a high ratio since they invest heavily in intellectual capital compared to their investment in plant, property, and equipment.

Another useful indicator of intellectual capital is the ratio of R&D expenditure to sales. Three of the companies listed in Table 2.3 spend more than 10% of sales for research and development. Given the R&D expenditure, only Microsoft and Cisco reap the reward of their high expenditure on R&D. We will use these four measures in our evaluation of companies in later chapters.

## Profit Models and Value Migration

Value migrates from outdated business designs to new ones that better satisfy customers and attract alliance partners. A business design has elements as listed in Table 2.4 [Slywotzky, 1997]. The **business design** results in a system for the business activities and relationships. As new economic discontinuities occur, companies redesign their business to match the new opportunities. Thus business is a dynamic process and any set of numbers that represents a company provides a snapshot when the investor needs a time series of pictures. Federal Express (FDX) pursued the opportunity for rapid package delivery and put together a business design based on a computerized logistics system. A business model can be developed using the four elements of Table 2.4. It is valuable to understand the basic business model utilized by the firm under study for investment. For example, the business design for Microsoft Windows is given in Table 2.5. Microsoft created a standard which resulted in increasing returns as the operating system Windows became the PC standard. This standard is the fundamental component around which the PC industry is built.

In order to protect a business profit, companies use a strategic control method.

**Table 2.4   The Four Key Elements of a Business Design**

| | |
|---|---|
| 1. Customer selection | — Which customers will receive real value from the product and allow the corporation to profit? |
| 2. Value capture | — How does the company profit? What is the profit model? |
| 3. Differentiation and strategic control | — How does the company protect the profit stream? |
| 4. Scope | — What activities does the company perform and what activities should be outsourced? |

**Table 2.5   The Microsoft Business Design for Windows**

| | |
|---|---|
| 1. Customer selection | — OEMs and personal computer manufacturers. Applications developers and users. |
| 2. Value capture | — Applications, upgrades, and increasing returns. |
| 3. Differentiation and strategic control | — Create the standard operating system. Work with applications developers. |
| 4. Scope | — Team development of software. Licensing. User and developer marketing. |

Table 2.6 lists the strategic control methods for profit protection [Slywotzky, 1997]. Very powerful companies enjoy a high profit index. A key strategic control factor is the recognition that a competitive strength is never powerful for long—making the ability to adapt to changes a crucial talent. We will seek to identify these powerful companies.

One example of a powerfully profitable company is Intel, which owns the standard for microprocessors and maintains the product lead time of at least one year. General Electric and Charles Schwab are examples of companies with excellent customer relationships and strong, consistent profitability. When you consider a candidate company for investment, try to determine its business model and its strategic control using Tables 2.4 and 2.6.

Powerful, productive companies dynamically adapt to a changing marketplace and continually improve their knowledge assets. Using the measure of sales per employee for four powerful companies, we show in Table 2.7 that great companies increase their intellectual capital over time. Note that these four companies approximately doubled their sales per employee over the nine-year period.

**Table 2.6   Strategic Control Design for Profit Protection**

| Profit Index | Design | Example |
|---|---|---|
| High | Own the standard | Microsoft |
| | Ability to adapt to market changes | Intel |
| | Manage the value chain | Coke |
| Medium | Build a customer relationship | GE |
| | Brand, copyright | New York Times |
| | Product lead time, first to market | Carnival Cruises |
| Low | Commodity with price advantage | Costco |
| | Commodity | Kellogg |

**Table 2.7   Increasing Intellectual Capital Productivity: Sales Per Employee ($000)**

| Company | 1991 | 1996 | 1999 |
|---|---|---|---|
| Dell | 550 | 630 | 750 |
| Hewlett-Packard | 195 | 359 | 403 |
| Merck | 231 | 421 | 555 |
| SBC Communications | 145 | 210 | 240 |

# 3 Risk and Reward

## This chapter covers

- Risk Defined
- Risk Factors
- Diversification
- Measures of Risk
- Time Horizon for Holding a Portfolio
- Risk versus Reward

## Risk Defined

An investor, in a portfolio of stocks, expects to achieve a reasonable return. This investor is also concerned with the probability that this return will be achieved and the possibility that the return will be less than as expected. We can consider risk as the potential for achieving less than the expected return or even an outright loss of the investment. Risk is the chance that the expected return will not materialize and that the equity securities may actually fall in price. Thus, we are interested in describing the variability of future returns. For example, if we purchase a one-year U.S. Treasury Bill with a fixed 4% interest rate, we know that if we hold this bond for one year we will receive our principal back plus the interest payments. This Treasury Bill is said to be risk-free. On the other hand, if we buy a stock in a small biotechnology or internet company, after one year the price of the stock may have gone up 30% or declined 30% or more. This is a risky stock. We define risk as follows:

**Risk** is a measure of the potential loss in the value of an investment as a result of changes in the condition of the economy between now and some future point in time.

**Regret** is the amount of self-insurance we can tolerate. It represents a limit beyond which we are unwilling to accept the consequences of a wrong choice. Regret depends on the investor circumstances. If you have invested $10,000, which is all your money, then you can probably accept a loss of $1,000. The risk-adjusted value of an investment is $U - \lambda R$ where $U$ = upside, $\lambda$ = risk-adjusted constant, usually greater than 1, and $R$ = downside or regret [Dembo, 1998]. The larger the value of $\lambda$, the more risk-averse is the investor. We are neutral if $\lambda = 1$, but mildly risk-averse if $\lambda = 2$. Thus, when $\lambda = 2$ we seek an upside at least twice the magnitude of the downside. Therefore, the value of an investment is zero if the upside is $1,000 and the downside is $500 when $\lambda = 2$.

If we use variance of returns to represent risk, we can examine the standard deviation of various investments based on past results. However, we recognize that the market is dynamic and the standard deviation may change over time. Nevertheless, standard deviation is a reasonable measure of the ride on the roller coaster the investor will probably endure. Standard deviation of returns is best used for a portfolio or a bundle of stocks or mutual funds.

If we select a money market fund for our investment we will experience low volatility of returns as well as low returns. Cash has no fluctuation and no return but is subject to a decline in value due to inflation. In the case of cash or a money market fund we must accept a low return (or none) but we attain low volatility of return. Cash or a money market fund is suitable for investment with short time horizons, say, one year or less.

Thus, the investor has a time horizon, an expected return, and a benchmark used to compare with the actual results. The usual assumption is that the changes in the prices of the stocks resemble a normal distribution and the benchmark is the S&P 500. Impressive evidence exists to show that changes in stock prices fall into a normal distribution. The total return for the S&P 500 for 1926–1995 (70 years) had a mean of 12.3% and a standard deviation of 20.5% [Bernstein, 1996]. (See Exhibit 3.1.) The individual investor is probably not going to hold one portfolio (the S&P 500) for 70 years. Consider the experience of the investor through the two-year decline of the S&P 500 in 1973–74 when the loss in equity values (including inflation) was 50%. Each investor asks: Can I withstand that drop and still stay in the market? If the investor purchased the S&P 500 equivalent on January 1, 1974, and held for five years, a positive return would have been achieved. Unfortunately, many investors sell their portfolio after a big drop, take their loss, and leave the stock market for good.

The investor seeks to minimize risk or losses. This can be achieved by adjusting the risk of the security portfolio, the length of the time horizon or the expected return. Investors think of expected return as a desirable outcome and the variance of return as an undesirable effect.

## Risk Factors

There are 10 risk factors that influence the overall risk of an investment as shown in Exhibit 3.2. Business risk means the company in which you invest may fail to meet the expected growth of revenues and income. Stock-specific risk is the risk of holding that

**Exhibit 3.1   Average Return and Volatility of Return, 1960–1997**

| | | Return, 1960–1997 | | |
| --- | --- | --- | --- | --- |
| | Annual Average | Worst Year | Best Year | Standard Deviation |
| **Investment** | (%) | (%) | (%) | (%) |
| S&P 500 | 11.6 | −26.5 | 37.4 | 20.5 |
| Long-term government bonds | 7.3 | −7.8 | 40.4 | 8.0 |
| International stocks (EAFE) | 12.3 | −23.2 | 69.9 | 22.1 |
| Money market fund | 7.0 | — | — | 5.0 |

**Exhibit 3.2   The 10 Factors of Risk**

1.   Business risk
2.   Stock-specific risk
3.   Liquidity risk
4.   Interest rate risk
5.   Market risk
6.   Inflation risk
7.   Political risk
8.   Taxation risk
9.   Leverage risk
10.  Valuation risk

one stock alone. You can protect yourself by purchasing several stocks and diversifying your portfolio.

Liquidity risk is the difficulty experienced when a stock has few buyers or sellers so that you may take a loss in order to sell. Interest rates affect all stocks since bonds and other securities may compete with your stock on the basis of the interest rate. When interest rates are rising, the value of stocks may decline. This decline may be particularly pronounced for bond-like securities such as high-yielding utility stocks.

Market risk is the risk associated with the movement of the overall market. If the entire market declines, as it did on October 19, 1987, the value of all stocks will decline. Inflation risk is the potential for loss of purchasing power over time. Historically, most stocks keep pace with inflation better than bonds or cash holdings. Political risk is prominent when investing in the stock of companies active in politically unstable areas. During the Asian economic crisis of 1998, the stocks of companies headquartered in Korea, Malaysia, Indonesia, and Hong Kong significantly dropped in value. The stocks of U.S. companies active in these countries also were impacted by the reduction of economic activity.

Taxation risk is the potential for a change in tax treatment of some companies in your portfolio. For example, if R&D tax credits were eliminated, companies heavily dependent upon research and development credits could be significantly impacted.

Leverage risk is the risk of loss due to high interest payments on a high level of debt. Typically, we will select companies with debt as a percent of capital less than 40%. Some industries, such as grocery stores, have higher debt/capital ratios but exhibit steady cash flows, thus reducing the impact of leverage.

Valuation risk is present for companies with a valuation ratio such as price to earnings (PE) or price to cash flow (PCF) significantly above the normal, historical range. We will seek to purchase stocks with reasonable valuation ratios.

## Diversification

In order to reduce the risk of loss, an investor uses diversification of security holdings. The idea with diversification is to combine a set of different companies over a range of size, location, and by industry. Market risk, stock-specific risk, business risk, political risk and taxation risk can be reduced by holding 20 or more stocks from different industries. Liquidity risk can be reduced by holding companies with revenues greater than $500 million. Inflation risk can be reduced by holding some stocks that will increase in value due to an increase in their holdings of natural resources (such as oil companies) or their holdings of commodities such as real estate. Leverage risk may be reduced by holding companies with low debt-to-capital ratios and a solid financial rating. Valuation risk can be reduced by selecting stocks with normal, historical price multipliers.

Because the stock prices of all the companies in your portfolio do not move in tandem, investment in a portfolio of 20 stocks is likely to be less risky than investment in one or two securities. The extent to which two securities move together is the covariance. The covariance between biotechnology stocks and utility stocks might be 0.3, indicating a relatively low linkage. A covariance of 1.0 indicates a perfect linkage.

We seek to build a portfolio of 20 to 30 stocks with a capitalization above $500 million. In addition, we select stocks from 10 attractive industries in order to seek a reduced covariance between stocks. Finally, we seek companies with relatively low business risk and leverage risk.

## Measures of Risk

We seek to minimize the risk of a portfolio of stocks. In order to select a stock to add to our portfolio we need to examine the riskiness of that stock. The report on a stock provided by the Value Line Survey will give us several indices of risk.

The Value Line **Safety Rating** is indicated for each stock. See Figure 3.1 for Item 1. The Safety Rating is a measure of the safety or lack of risk. Safety is ranked from 1 (very safe), to 5 (relatively unsafe). Stocks with a rank of 1 or 2 hold their price relative to the market. If an investor seeks safety (low risk), one might select stocks with a rating of 1 or 2. As shown in Table 3.1, the decline experienced by a group of stocks with a 1 rating fell only –6.1% in the 1998 decline, compared to –28.5% for the Value Line Arithmetic Average (Value Line, 1999). High Safety Ranks are associated with stocks that are less volatile than the average and with generally strong balance sheets.

Another measure of risk is **beta,** which is a measure of the covariance between a security return and the market return scaled by the variance of the market return. Thus, beta is the change in a securities return divided by the accompanying change in the return to the market portfolio. Traditional estimates rely on regression analysis to estimate beta. Beta is a measure of the sensitivity of a stock to market movements. The

**Table 3.1    Value Line Safety Rank and Decline for Mid-1998**

| Group of Stocks with Safety Rating | Decline Percentage (%) for 4/22/98–10/08/98 |
|---|---|
| 1 | –6.1 |
| 2 | –14.0 |
| 3 | –29.7 |
| 4 | –41.7 |
| 5 | –37.8 |
| S&P 500 | –15.1 |
| Value Line Arithmetic Average | –28.5 |

beta of the market is 1.0 and a stock to market movements. The beta of the market is 1.0, and a stock with a beta equal to 1.5 implies that when the market declines by 20%, the stock is expected to decline by 30%. The beta of WalMart is 1.00 as shown by Item 2 in Figure 3.1. The beta of Yahoo is 2.00, indicating a very volatile stock.

The measure beta has not been very successful in explaining the differences in returns of individual stocks, and the link between beta and return has been modest. Nevertheless, beta is often useful as one of several measures of risk.

Another measure of risk provided by Value Line is **Stock Price Stability,** which is a measure of price volatility. The stability index ranges from 5 (lowest stability or highest volatility) to 100 (highest stability or lowest volatility). We usually state risk as the historical volatility of return. However, it is important to recognize that risk is a forward-looking concept while volatility is a historical statistic. The Stock Price Stability for WalMart is 70 as shown by Item 3 in Figure 3.1.

In Table 3.2 we show the Value Line Safety Rating, the beta, and the Stock Price Stability for 10 stocks. All three measures of risk provide similar information regarding the safety (risk) of an individual stock. The Stock Price Stability measure appears to be the most sensitive to volatility of price (or return). We will use beta and the Stock Price Stability rating as measures of a stock's risk, but emphasize the Stock Price Stability.

## Time Horizon for Holding a Portfolio

Numerous studies have shown that the variability of return of a portfolio may be reduced by increasing the holding period. As shown in Table 3.3, if you are prepared to hold a portfolio of 10 years or more, the variability of return can be significantly reduced.

The volatility of a portfolio of stocks such as the S&P 500 Index can be quite high in a one-year period as shown in Figure 3.2. A decline of 22% occurred in Fall Quarter

**Table 3.2   Risk Measures for 10 Stocks**

| Company | Value Line Safety Rating | Beta | Stock Price Stability | Annual Return 1993–1998 (%) |
|---|---|---|---|---|
| Otter Tail Power | 2 | 0.55 | 95 | 6.3 |
| New Plan Excel | 2 | 0.85 | 95 | 28.7 |
| New Jersey Resources | 2 | 0.55 | 90 | 14.8 |
| Alltel | 2 | 0.85 | 85 | 12.0 |
| Walgreen | 2 | 1.10 | 70 | 33.3 |
| State Street | 3 | 1.25 | 60 | 28.2 |
| Medtronic | 3 | 1.10 | 60 | 48.5 |
| Lilly | 3 | 1.10 | 55 | 46.6 |
| Intel | 3 | 1.00 | 35 | 50.6 |
| Schwab | 3 | 1.85 | 25 | 64.7 |
| Cisco | 3 | 1.50 | 20 | 66.8 |
| EMC Corp. | 3 | 1.50 | 15 | 59.4 |
| Uniphase | 4 | 1.75 | 5 | 98.5 |

# WAL-MART STORES NYSE-WMT

| | | |
|---|---|---|
| RECENT PRICE | 47 | |
| P/E RATIO | 40.2 | (Trailing: 47.5 / Median: 24.0) |
| RELATIVE P/E RATIO | 2.26 | |
| DIV'D YLD | 0.4% | |

| | High | Low |
|---|---|---|
| TIMELINESS 2 Lowered 5/21/99 | 4.2 | 3.0 |
| SAFETY 2 New 7/27/92 | 5.6 | 3.8 |
| TECHNICAL 2 Lowered 3/19/99 | | |
| BETA 1.00 (1.00 = Market) | | |

LEGENDS
18.0 x "Cash Flow" p sh
.... Relative Price Strength
2-for-1 split 7/87
2-for-1 split 7/90
2-for-1 split 7/90
2-for-1 split 2/93
2-for-1 split 4/99
Options: Yes
Shaded area indicates recession

## 2002-04 PROJECTIONS

| | Price | Gain | Ann'l Total Return |
|---|---|---|---|
| High | 75 | (+60%) | 13% |
| Low | 55 | (+15%) | 5% |

### Insider Decisions

| | J | J | A | S | O | N | D | J | F |
|---|---|---|---|---|---|---|---|---|---|
| to Buy | 1 | 0 | 0 | 2 | 0 | 0 | 1 | 0 | 0 |
| Options | 2 | 0 | 0 | 2 | 0 | 0 | 1 | 0 | 1 |
| to Sell | 4 | 1 | 0 | 1 | 1 | 2 | 6 | 3 | 2 |

### Institutional Decisions

| | 2Q1998 | 3Q1998 | 4Q1998 |
|---|---|---|---|
| to Buy | 334 | 376 | 441 |
| to Sell | 319 | 332 | 328 |
| Hld's(000) | 1692757 | 1697381 | 1696342 |

| | 1983 | 1984 | 1985 | 1986 | 1987 | 1988 | 1989 | 1990 | 1991 | 1992 | 1993 | 1994 | 1995 | 1996 | 1997 | 1998 | 1999 | 2000 | © VALUE LINE PUB., INC. | 02-04 |
|---|---|---|---|---|---|---|---|---|---|---|---|---|---|---|---|---|---|---|---|---|
| Sales per sh A | 1.04 | 1.43 | 1.88 | 2.64 | 3.53 | 4.56 | 5.70 | 7.14 | 9.55 | 12.06 | 14.65 | 17.96 | 20.42 | 22.87 | 26.32 | 30.71 | 36.05 | 41.65 | | 62.80 |
| "Cash Flow" per sh | .05 | .08 | .09 | .13 | .18 | .23 | .30 | .36 | .45 | .57 | .69 | .82 | .88 | .99 | 1.15 | 1.41 | 1.75 | 2.05 | | 3.35 |
| Earnings per sh B | .04 | .06 | .07 | .10 | .14 | .19 | .24 | .29 | .35 | .44 | .51 | .59 | .60 | .67 | .78 | .99 | 1.25 | 1.50 | | 2.45 |
| Div'ds Decl'd per sh ■C | .00 | .01 | .01 | .01 | .02 | .02 | .03 | .04 | .04 | .05 | .07 | .09 | .10 | .11 | .14 | .16 | .20 | .25 | | .45 |
| Book Value per sh | .16 | .22 | .28 | .37 | .50 | .66 | .88 | 1.17 | 1.52 | 1.90 | 2.34 | 2.77 | 3.22 | 3.74 | 4.13 | 4.71 | 5.30 | 6.00 | | 9.40 |
| Common Shs Outst'g D | 4477.3 | 4487.1 | 4496.7 | 4514.9 | 4520.9 | 4524.7 | 4529.1 | 4569.1 | 4596.1 | 4599.3 | 4597.5 | 4594.0 | 4586.0 | 4586.0 | 4482.0 | 4482.0 | 4440.0 | 4400.0 | | 4300.0 |
| Avg Ann'l P/E Ratio | 26.1 | 20.4 | 22.5 | 27.8 | 27.7 | 20.3 | 20.7 | 24.4 | 32.8 | 32.8 | 27.0 | 20.9 | 20.4 | 18.4 | 21.8 | 31.2 | Bold figures are Value Line estimates | | | 27.0 |
| Relative P/E Ratio | 2.21 | 1.90 | 1.83 | 1.89 | 1.85 | 1.69 | 1.57 | 1.81 | 2.10 | 1.99 | 1.59 | 1.37 | 1.37 | 1.15 | 1.26 | 1.64 | | | | 1.80 |
| Avg Ann'l Div'd Yield | .4% | .5% | .5% | .4% | .4% | .5% | .6% | .5% | .4% | .4% | .5% | .7% | .8% | .9% | .8% | .5% | | | | .7% |

| | 1983 | 1984 | 1985 | 1986 | 1987 | 1988 | 1989 | 1990 | 1991 | 1992 | 1993 | 1994 | 1995 | 1996 | 1997 | 1998 | 1999 | 2000 | | 02-04 |
|---|---|---|---|---|---|---|---|---|---|---|---|---|---|---|---|---|---|---|---|---|
| Sales ($mill) A | 25611 | | | | | | 32602 | 43887 | 55484 | 67345 | 82494 | 93627 | 104859 | 117958 | 137634 | 160000 | 183200 | | | 270000 |
| Gross Margin | 23.3% | | | | | | 22.8% | 21.8% | 21.6% | 21.9% | 21.8% | 21.8% | 21.6% | 22.2% | 22.4% | 22.5% | 22.5% | | | 22.5% |
| Operating Margin | 7.5% | | | | | | 7.0% | 6.6% | 6.6% | 6.6% | 6.2% | 5.8% | 5.6% | 5.8% | 6.1% | 6.3% | 6.5% | | | 6.7% |
| Number of Stores | 1525 | | | | | | 1721 | 1932 | 2136 | 2440 | 2759 | 2943 | 3054 | 3406 | 3599 | 3790 | 4000 | | | 4700 |
| Net Profit ($mill) | 1075.9 | | | | | | 1291.0 | 1608.5 | 1994.8 | 2333.3 | 2681.0 | 2740.0 | 3056.0 | 3526.0 | 4430.0 | 5575 | 6625 | | | 10600 |
| Income Tax Rate | 37.0% | | | | | | 36.8% | 37.0% | 37.0% | 36.8% | 37.1% | 37.0% | 37.0% | 37.0% | 37.4% | 37.5% | 37.5% | | | 37.5% |
| Net Profit Margin | 4.2% | | | | | | 4.0% | 3.7% | 3.6% | 3.5% | 3.3% | 2.9% | 2.9% | 3.0% | 3.2% | 3.4% | 3.6% | | | 3.9% |
| Working Cap'l ($mill) | 1867.3 | | | | | | 2424.4 | 3571.6 | 3443.3 | 4708.4 | 5365.0 | 5877.0 | 7036.0 | 4892.0 | 4370.0 | 4600 | 4800 | | | 5500 |
| Long-Term Debt ($mill) | 1272.6 | | | | | | 1898.9 | 3277.9 | 4845.0 | 7960.2 | 9709.0 | 10600 | 10016 | 9674.0 | 9607.0 | 9600 | 9500 | | | 9100 |
| Shr. Equity ($mill) | 3965.6 | | | | | | 5365.5 | 6989.7 | 8759.2 | 10752 | 12726 | 14756 | 17143 | 18503 | 21112 | 23500 | 26500 | | | 40500 |
| Return on Total Cap'l | 21.7% | | | | | | 18.8% | 16.8% | 15.8% | 13.3% | 13.4% | 12.2% | 12.6% | 13.8% | 15.6% | 18.0% | 19.0% | | | 22.0% |
| Return on Shr. Equity | 27.1% | | | | | | 24.1% | 23.0% | 22.8% | 21.7% | 21.1% | 18.6% | 17.8% | 19.1% | 21.0% | 23.5% | 25.0% | | | 26.0% |
| Retained to Com Eq | 24.0% | | | | | | 21.1% | 20.2% | 20.0% | 18.9% | 18.0% | 15.5% | 15.0% | 15.8% | 17.7% | 19.5% | 21.0% | | | 21.0% |
| All Div'ds to Net Prof | 12% | | | | | | 12% | 12% | 12% | 13% | 13% | 15% | 16% | 16% | 16% | 16% | 17% | | | 18% |

| Percent shares traded | | | | |
|---|---|---|---|---|
| | 6.0 | | | |
| | 4.0 | | | |
| | 2.0 | | | |

## Target Price Range

| | 2002 | 2003 | 2004 |
|---|---|---|---|
| | | | 100 |
| | | | 80 |
| | | | 64 |
| | | | 48 |
| | | | 40 |
| | | | 32 |
| | | | 24 |
| | | | 20 |
| | | | 16 |
| | | | 12 |
| | | | 8 |
| | | | 6 |

### % TOT. RETURN 4/99

| | THIS STOCK | VL ARITH. INDEX |
|---|---|---|
| 1 yr. | 82.9 | 0.0 |
| 3 yr. | 293.4 | 57.7 |
| 5 yr. | 277.8 | 119.0 |

## CAPITAL STRUCTURE as of 1/31/99

Total Debt $10613 mill. Due in 5 Yrs $4334 mill.
LT Debt $9607 mill. LT Interest $730.0 mill.
Incl. $2699 mill. capitalized leases.
(LT interest earned: 11.4x; total interest
coverage: 10.2x)     (31% of Cap'l)

Leases, Uncapitalized Annual rentals $394 mill.

Pension Liability None

Pfd Stock None

Common Stock 4,482,000,000 shs. (69% of Cap'l)
MARKET CAP: $210 billion (Large Cap)

3

BUSINESS: Wal-Mart Stores, Inc. is the world's largest retailer, operating about 1,870 discount stores, 565 supercenters (includes sizable grocery departments), and 450 Sam's Clubs in the U.S., plus 870 foreign stores, mainly in Latin America and Canada. Most stores are leased and are kept within a 400-mile radius of one of the expanding network of distribution centers. Also distributes goods to convenience stores through its McLane unit. Total store space: about 385 mill. sq. ft.; sales per square foot in the U.S. about $385; international, $290. Has 910,000 employees. Officers & Dirs. own 40.3% of shares (4/99 proxy). Chrmn.: S. Robson Walton. Pres.: David D. Glass. Inc.: DE. Address: Box 116, Bentonville, AR 72716. Tel.: 501-273-4000. Internet: www.wal-mart.com.

**Wal-Mart is progressing in both of its major growth endeavors.** First, margins at the U.S. supercenter segment (almost 35% of domestic selling space) are approaching the level of the discount stores. This development reflects ongoing economies of scale, large year-to-year sales increases of private-label grocery items, and expense-leveraging benefits of same-store sales gains of about 9%, on average, in recent quarters. The total number of domestic supercenters will likely more than double, to somewhat over 1,200 units, during the next four years. Second, the foreign store count is slated to expand by about 15% both this year and next. (Fiscal years end on January 31st.) Thanks to this internal growth, as well as to an ongoing acquisition program, management expects increases in international sales and profits to account for 25%–30% of Wal-Mart's projected top- and bottom-line gains over the coming 3 to 5 years.

**We look for earnings to advance by about 25% and 20% in fiscal 1999 and 2000, respectively.** Same-store sales gains at the domestic supercenters exceeded 10% in fiscal 1999's first quarter, thanks, in part, to more price rollbacks than usual. Despite these cuts, margins at this division improved because of the leveraging effect of increases in both customer count per store and average transaction size. The addition of automobile-related services, among others, and store-layout revisions are other factors that augur well for solid profit gains at this segment during the coming quarters. Moreover, **Wal-Mart's prospects for solid share-net advances through 2002–2004 are bright.** Its planned rate of annual supercenter openings, which increased by 20%, to 150, in the latter half of fiscal 1998, far exceeds the likely additions of such facilities by the handful of domestic competitors that operate in this arena. Also, we expect WMT's profits in European and Asian markets to ramp up rapidly during this time frame. Finally, we look for a gradual reduction in shares outstanding and long-term debt.

**Wal-Mart stock is timely.** This issue, which represents over 50% of the market capitalization of our Retail Industry group, should be a core holding.

*David R. Cohen*

*May 21, 1999*

| CURRENT POSITION ($MILL.) | 1996 | 1997 | 1/31/99 |
|---|---|---|---|
| Cash Assets | 883 | 1447 | 1879 |
| Receivables | 845 | 976 | 1118 |
| Inventory (LIFO) | 15897 | 16497 | 17076 |
| Other | 368 | 432 | 1059 |
| Current Assets | 17993 | 19352 | 21132 |
| Accts Payable | 7628 | 9126 | 10257 |
| Debt Due | 618 | 1141 | 1006 |
| Other | 2711 | 4193 | 5499 |
| Current Liab. | 10957 | 14460 | 16762 |

| ANNUAL RATES of change (per sh) | Past 10 Yrs. | Past 5 Yrs. | Est'd '96-'98 to '02-'04 |
|---|---|---|---|
| Sales | 22.0% | 17.0% | 15.5% |
| "Cash Flow" | 21.0% | 15.5% | 19.0% |
| Earnings | 19.0% | 13.5% | 20.0% |
| Dividends | 24.0% | 20.0% | 23.0% |
| Book Value | 23.5% | 17.0% | 14.5% |

| Fiscal Year Begins | QUARTERLY SALES ($ mill.) A | | | | Full Fiscal Year |
|---|---|---|---|---|---|
| | Apr.30 | Jul.31 | Oct.31 | Jan.31 | |
| 1996 | 22772 | 25587 | 25644 | 30856 | 104859 |
| 1997 | 25409 | 28386 | 28777 | 35386 | 117958 |
| 1998 | 29819 | 33521 | 33509 | 40785 | 137634 |
| 1999 | 34700 | 39000 | 38900 | 47400 | 160000 |
| 2000 | 39800 | 44750 | 44600 | 54050 | 183200 |

| Fiscal Year Begins | EARNINGS PER SHARE A B | | | | Full Fiscal Year |
|---|---|---|---|---|---|
| | Apr.30 | Jul.31 | Oct.31 | Jan.31 | |
| 1996 | .13 | .15 | .15 | .24 | .67 |
| 1997 | .15 | .17 | .18 | .28 | .78 |
| 1998 | .18 | .23 | .23 | .35 | .99 |
| 1999 | .25 | .28 | .29 | .43 | 1.25 |
| 2000 | .30 | .34 | .35 | .51 | 1.50 |

| Calendar | QUARTERLY DIVIDENDS PAID ■ C | | | | Full Year |
|---|---|---|---|---|---|
| | Mar.31 | Jun.30 | Sep.30 | Dec.31 | |
| 1995 | .022 | .025 | .025 | .025 | .10 |
| 1996 | .025 | .027 | .027 | .027 | .11 |
| 1997 | .027 | .034 | .034 | .034 | .13 |
| 1998 | .0345 | .039 | .039 | .039 | .15 |
| 1999 | .039 | .05 | | | |

(A) Fiscal year ends Jan. 31st of following calendar year. Sales exclude rentals from licensed depts.
(B) Based on diluted shares. Next earnings report due mid-Aug.
(C) Next dividend meeting about June 1. Goes ex about June 12. Dividend payment dates: Jan. 5, April 10, July 8, Oct. 3.
■ Dividend reinvestment plan available.
(D) In millions, adjusted for stock splits.

| Company's Financial Strength | A+ |
|---|---|
| Stock's Price Stability | 70 |
| Price Growth Persistence | 45 |
| Earnings Predictability | 90 |

**To subscribe call 1-800-833-0046.**

Figure 3.1  The Value Line Survey Page for WalMart Stores (Courtesy of Value Line Survey)

**Table 3.3    Effect of Holding Period, S&P 500, 1926–97**

| Holding Period | 1 Year | 3 Years | 5 years | 10 Years | 20 Years |
|---|---|---|---|---|---|
| Percentage (%) of Holding Periods that Produced Losses | 27 | 13 | 10 | 3 | 0 |

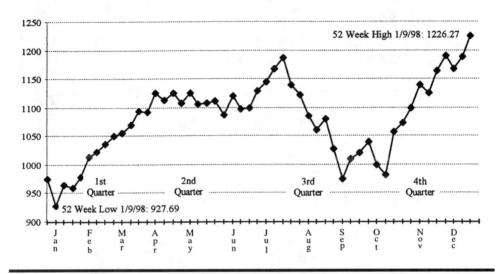

**Figure 3.2    The S&P 500 Index for 1998 (Source: Safeco Funds)**

of 1998. Clearly, it is wise to select a holding period of at least five years and avoid a panic sale on the occasion of a big drop as happened in the Fall Quarter of 1998.

The probability of achieving an expected return for a portfolio can be influenced by the holding period. Table 3.4 shows an approximate probability of achieving an expected average annual return for selected portfolios and the estimated required holding period to achieve the expected return with the estimated probability.

**Table 3.4    The Holding Period for Selected Portfolios**

| Portfolio | Probability | Expected Average Annual Return (%) | Required Holding Period (Years) |
|---|---|---|---|
| Treasury Bills | 1.00 | 4.5 | 1 |
| 10-Year Treasury Bonds | 0.90 | 6.0 | 2 |
| Utility Portfolio | 0.80 | 12.0 | 3 |
| Growth Portfolio | 0.70 | 25.0 | 5 |
| High Growth Portfolio | 0.60 | 35.0 | 6 |
| Speculative Small Caps | 0.50 | 40.0 | 7 |

# Risk versus Reward

Reaching for high returns inevitably carries high risk. There is a chance of failure or loss even for long-term investors. If you hold a diversified portfolio of stocks for 20 years and don't panic in a prolonged downturn, you are most likely to achieve reasonable returns. High inflation is the biggest risk for stocks, and most investors appreciate the fight against inflation waged by the Federal Reserve Bank. Hopefully, the Fed can restrain inflation to remain below 2.5%. Then we can seek to build a portfolio that provides an average annual return of 11 to 15%. A company that can annually return 10% nine out of 10 years is less risky than a company that can potentially return 16% seven out of 10 years. They both may be good investments if you can accept the lower return of the first stock and the higher risk of the second stock. If you place both of them in your portfolio, you will benefit from the diversification.

Most readers would agree that investors should be compensated for taking on increased risk. Stock prices and thus returns should adjust to offer higher returns when more risk is assumed or perceived by investors. Risk is the other side of opportunity. To achieve big gains, we have to take big risks. Support for beta as the measure of risk has waivered as of late.

We can draw a risk-versus-reward model as shown in Figure 3.3. We have assigned places on the risk-versus-reward line for different types of stocks. Note we use a general designation of risk and do designate it as one parameter, such as beta or standard deviation of returns. Returning to Table 3.2 for ten stocks (an unscientific sample) we

## Risk versus Reward

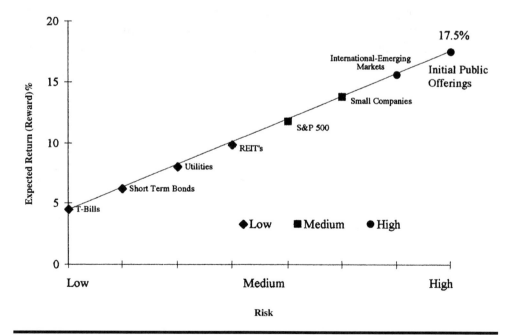

**Figure 3.3    Risk versus Reward**

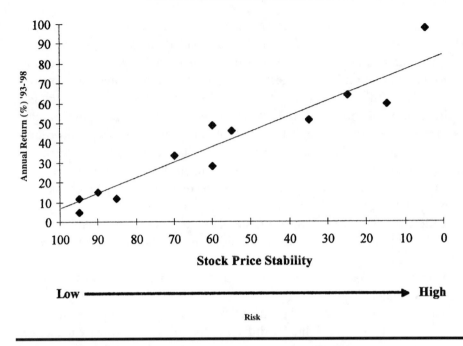

**Figure 3.4  Risk Versus Reward Ten Selected Stocks, 1993–98**

plot actual annual return versus Stock Price Stability where 100 is the highest stability or the lowest risk as shown in Figure 3.4. We can estimate a linear regression line as shown and note a reasonably linear relationship for risk versus reward. While not exact or scientifically proven, a reasonable estimate of risk is provided by the Value Line Stock Price Stability. Thus, we select it as one of our best risk measures for selecting stocks.

# 4 | An Investment Plan

## This chapter covers

- Build and Keep an Investment Plan
- Expected Return and Time Horizon
- Risk and Timing—Buy and Hold
- Investment Cash Availability
- Market Capitalization of Stocks
- Powerful, Dominant Companies
- Dividends and Stock Repurchases

## Build and Keep an Investment Plan

The key to successful investing is building an investment plan that fits you and then sticking to it. Of course, during the building phase you may try several experiments to test your risk profile and assess the time you are willing to put into maintaining the investment process. Nevertheless, the goal is to build a plan that is rational and maintainable. The key word is conviction. The process of building a plan is summarized in Exhibit 4.1.

The common elements of a stock investment plan include the expected return, acceptable risk, probable trading pattern, and planned time horizon. The first step is to establish your specific goals and the second step is to set standards or ranges for risk and reward that are acceptable. Then, as illustrated in Chapter 1, develop scenarios for the future economy as a background for stock selection. Select a buying pattern and a

**Exhibit 4.1   The Process of Building an Investment Plan**

**Step**

1. Define specific investment goals in terms of expected return, risk, trading, and time horizon.
2. Set standards for Risk and Return in terms of acceptable ranges.
3. Develop scenarios for the future economy.
4. Select a buying plan and a holding plan.
5. Select candidate stocks.
6. Take action with several attractive stocks.
7. Test the results of step 6 and adjust the parameters of Steps 1–5 as necessary.

**Exhibit 4.2   Factors of an Investment Plan**

- Expected return
- Ultimate time horizon and holding period
- Acceptable risk = volatility of return
- Trading and timing issues
- Investment cash availability and timing
- Diversification strategy

**Exhibit 4.3   An Example of the Factors of an Investment Plan**

- Expected return = 14% Average per Year
- Time Horizon = 10 Years
- Acceptable Risk = Medium, Market Level
- Trading and Timing = Buy and Hold, Adjust Slowly
- Investment Cash = Purchase Stocks Quarterly with New Cash
- Diversification = 30 Stocks in 15 Industries

holding plan for the portfolio. Then select candidate stocks using appropriate criteria. Take action by purchasing several attractive stocks and then test the results. Using the test results, adjust the elements of Steps 1–6. When you develop a well-reasoned, disciplined investment plan you will be able to stick to it and adjust it slowly as results dictate. As the economy changes and interest rates, economic growth and inflation change then use your new cash to invest in companies that are appropriate for the changing conditions.

The six factors of an investment plan are shown in Exhibit 4.2. An illustrative example of one investor's six factors are given in Exhibit 4.3. This example shows the investment plan for a long-term investor with a primary buy and hold strategy and 30 stocks in 15 industries. This investor is willing to accept a portfolio volatility equivalent to that of the stock market as represented by the S&P 500.

## Expected Return and Time Horizon

What is a reasonable rate of return on a portfolio of stocks over a ten-year period? Table 4.1 shows the return of the S&P 500 stocks for the period 1989–1998. The compound annual return during that period was 19.2% where the return varied between −3.1% and +37.6%. An expected return for a 10-year period normally might be 14%. For a 10-year holding period the range of return might be 11% at the low end (one standard deviation) and 17% at the high end. The standard deviation for stocks has historically been about 21%.

**Table 4.1   Annual Total Return of the S&P 500 for 1989–1998**

| Year | '89 | '90 | '91 | '92 | '93 | '94 | '95 | '96 | '97 | '98 |
|---|---|---|---|---|---|---|---|---|---|---|
| Total Return (%) | 31.7 | −3.1 | 30.5 | 7.6 | 10.1 | 1.3 | 37.6 | 23.0 | 33.4 | 28.6 |

## Risk and Timing—Buy and Hold

We seek to minimize risk to our portfolio, which is achieved by holding the portfolio for at least 10 years. It is important to hold the stocks for the ten years and avoid excessive trading in and out of the market. Frequent buying and selling is expensive, and it is very difficult if not impossible to efficiently time the market. Trading brings capital gains taxes and to be effective requires buying shares before they rise and selling before they fall. For most investors market timing is notoriously difficult. Market corrections are frequent, but it is very difficult to know if they are major declines. The most common cause of major bear market declines are interest-rate hikes by the Federal Reserve Bank. These rate changes can not be reliably anticipated. Overvaluation of the stock market is not a reliable predictor of a major decline. Furthermore, as an investor, you hold 20 or 30 selected stocks—not the market.

The average investor tilts his/her stock portfolio towards higher-risk, smaller capitalization stocks and turns over 80 percent of its portfolio annually. The net returns after taxes and transaction costs of this strategy lag the returns of an investor who buys and holds a broad-based market index [Barber, 1999]. During the past century, the American stock market has far outperformed the bond market and the foreign markets. Can we expect that stocks will outperform over the next 10 or 15 years? We can expect significant returns over the next 10 years if we buy and hold the best, dominant companies in the U.S. and elsewhere.

## Investment Cash Availability

Most investors have cash available periodically for investment. Investors are wise to make additional investments every month or every quarter. Regular periodic investment of a fixed amount is called **dollar cost averaging.** This method helps provide a disciplined approach as an alternative to trying to decide when the best value is available. If you invested $250 monthly in shares of McDonald's for the 10 years January 1, 1986, to December 31, 1995, you would have invested $30,000. The value of your shares at the end of 1995 would have been $86,231. A regular investment plan helps most investors achieve their goal over the long-term. The easiest and most effective plan is to keep investing and avoid the anxiety of trying to outguess the timing of market fluctuations.

## Market Capitalization of Stocks

Financial research argues that there are three stock market factors: a firm's size, price-to-book ratio, and the firm's beta [Fama, 1993]. Beta is unable to consistently explain observed differences in return. Over the long term, there is little to distinguish firm size, price to book, price momentum or earnings momentum as asset pricing factors. The average annual total return for selected market indices for the period 1989–98 is given in Table 4.2. The return for large and mid cap stocks outperformed the return for small cap stocks during that period. Only investments in firms with low price-to-book ratios provide a positive margin of return compared to the market. This low

**Table 4.2   Average Annual Total Return for Selected Market Indices, 1989–98**

| Type | Index | Total Return (%) |
|---|---|---|
| Large Cap | S&P 500 | 19.47 |
| Technology | Pacific Stock Exchange Technology | 23.94 |
| Small Cap | Russell 2000 | 13.88 |
| Global | MS World | 11.81 |

MS: Morgan Stanley
Cap: Capitalization

price-to-book method can be described as a contrarian strategy and is described in Chapter 11. We will emphasize large and mid-cap stocks.

Since we seek to reduce trading costs and tax consequences, we have chosen a modified buy and hold strategy. This investment strategy is served by selecting only mid-capitalization and large-capitalization stocks. **Market capitalization** is the current market price of one share of a company's common stock multiplied by the number of shares outstanding. A large stock capitalization strategy affords low costs of trading, liquidity, and low turnover. In addition, large companies outperform because they can better cope with disinflation. In general, we select companies with a market capitalization of $1 billion or greater. Value Line provides the market capitalization of each stock on the standard review page of each firm. As an example, see Item 4 of Figure 3.1 which shows a market capitalization for Wal-Mart of $210 billion.

We do not preclude holding a few companies with smaller capitalizations in the range of $500 million to $1 billion if they hold attractive advantages in their industry. In recent years the stock market has favored larger capitalization stocks. Large cap companies have generated better profit growth in the period '93–'99 and are the beneficiaries of disinflation. An example of a mid-cap stock with a market capitalization of about $750 million is Philadelphia Suburban (PSC), a water utility company. This is a relatively low risk stock with an attractive, steady return of about 14% annually. The stock price stability of PSC is 90 and the beta is 0.50. Thus, if the market was to decline 20% we would expect PSC to only decline 10% or less.

## Powerful, Dominant Companies

Companies with revenues over $1 billion are in a position, if well managed, to generate profits and cash flow and build the resources necessary to adapt to the challenges of the future. Stocks that deliver steady growth of earnings and revenues are valued highly by investors. Their earnings are more reliable than those of most other stocks. For the three-year period 1995–98, the S&P 500 (larger stocks) returned an average annual return of 23.8%. By contrast the Russell 2000 Index (small cap) returned 11.6% annually. We seek to purchase larger **blue chip** companies. A blue chip company sells at a higher price because of public confidence in its long record of steady earnings and the reasoned expectation that this steady stream can continue for the next five or more years. These companies have a de facto franchise. They lead their industries through the power of their products, distribution systems, brand names, innovation, and the efficiency of their technology. They often have little competition, strong

**Exhibit 4.4   Elements of Great Companies**

- Innovation
- Intellectual capital
- Ability to adjust
- Quality products
- Customer satisfaction
- Lower cost structure
- Strong distribution channels
- Intellectual leadership

financial positions and agile, intelligent leadership. The elements of great franchise companies are shown in Exhibit 4.4.

To a great extent, the S&P 500 stocks represent the best of the American economy. Companies added to the S&P 500 are chosen from leadership companies in leadership industries. For example, in 1998 Safeway, Firstar, and America OnLine replaced Chrysler and Amoco.

The blue chip companies boast a large market share and the ability to expand profit margins. Blue chip firms have a sustainable competitive advantage such as a secret formula or a computer software standard. These are the characteristics associated with near-monopolies like Microsoft. The large global companies use this size to spread their costs, hold down prices and fend off competitors. The franchise is like a barrier to entry since there is a huge cost to establishing it. Coke, for example, owes its 50% share of the world soft drink market to its well-developed brand and distribution network. Pharmaceutical companies, such as Merck or Pfizer, have patents on their products. Intel has an 84% share of the market for microprocessors and a well-established brand (Intel Inside) monopoly.

## Dividends and Stock Repurchases

A **dividend** is a payment made to shareholders, usually in cash but sometimes granted as additional shares. Companies that pay regular cash dividends traditionally issue them quarterly. A stock's current yield is the stock's annual per-share dividend divided by its current market price. Firms paying dividends attract investors that seek some current yield while holding the stock for long-term capital gains. Most companies are investing in internal growth opportunities and have cut back on dividend increases. Nevertheless, dividend increases are often interpreted by investors as a positive signal of future success. Stocks with dividends are an excellent alternative to holding bonds for income. Utility stocks and REITs provide a yield of 4% to 6%. If the investor decides to reinvest the dividends in a company's dividend reinvestment plan, the investor has opted for a form of dollar-cost averaging.

When dividends are paid, it is costly to shareholders because they must pay taxes on them. As an alternative, many firms use their funds to repurchase their own shares rather than issue dividends. The dividend yield of the S&P 500 has declined steadily over the past decade and is less than 1.5%. During the same period stock repurchases have risen from $20 billion in 1990 to $180 billion in 1998. Dividends in 1998 amounted to $250 billion.

Dividends provide a ready source of cash to investors with some income needs and a source of downside stability for a stock. Furthermore, stocks of high quality, blue chip companies typically regularly increase their modest dividends payments. For example, Abbott Laboratories yields about 1.4% but has increased the cash dividend 318% over the period 1988–98. In general, consistently rising dividends correlate well with consistently rising earnings.

High-yielding stocks like utilities and REITs are more decline resistant due to their yield. For example, consider a stock yielding 5.5% if the market drops 30%. If the utility stock drops 30% it would yield 7.9%. This yield would probably exceed corporate bond yields. Thus, the utilities normally only drop about one-half of the decline of the S&P 500. If we experience a prolonged period of declining or low returns, the dividend will again become more important to investors.

The benefits of corporate repurchases of their own stock are shown in Exhibit 4.5. Buyback programs usually help a company's share price by reducing the number of shares available, thus increasing the earnings per share. For the period 1978–1997, stocks with buyback programs provided an annual return of 19.7% compared to 15.5% for the average stock. Buyback programs make sense for companies with high cash flows and an undervalued stock price. U.S. corporations announced a record 1,962 buyback programs in 1998.

In the method developed in this book, we use stocks with relatively high dividends as an alternative to corporate bonds. Also, we identify steady growth stocks with growing dividends to serve as stable, blue chip holdings.

**Exhibit 4.5   Dividends and Share Repurchases**

| Dividends | Repurchases |
| --- | --- |
| **Advantages:** | |
| ■ Provides source of cash income to shareholders | ■ Increases remaining shareholders percentage ownership |
| ■ Sends signal of good future earnings | ■ Boosts return on equity |
| ■ Alternative to Bonds | ■ Increases earnings per share |
| ■ Cushions decline of share price | ■ No immediate tax consequences |
| ■ Can provide a return in a flat market | ■ Sends signal that company thinks the shares are undervalued |
| **Disadvantages:** | |
| ■ Uses capital that could be used for new initiatives | ■ Uses capital that could be used for new initiatives |
| ■ Taxable income to shareholder | |

# 5 | The Best Diversified Stock Portfolio

## This chapter covers

- Growth versus Value Investing
- Sectors and Industries
- Asset Allocation
- Building a Diversified Stock Portfolio Using Four Subportfolios
- Selecting Your Best Total Stock Portfolio

## Growth versus Value Investing

**W**riters and speakers often discuss value and growth investing as two contrasting styles. Growth investing is concerned with higher growth companies while value investors are concerned with undervalued companies. We only are interested in companies exhibiting some growth of revenues and earnings. For growth stocks we seek growth greater than that of the average stock in the Value Line Survey or the S&P 500. Typically a growth stock will have an earning growth rate of 14% to 16%. Value stocks may have a growth rate of earnings between 6% and 14%. Stocks with a current negative or zero growth rate may be candidates for our contrarian portfolio if we can see that growth rate turning positive.

We can gain insight into the two styles of investing by looking at an approximate formula for the one-year return on a stock as:

$$\text{Return} = \text{Dividend Yield} + \Delta E + \Delta PE$$

where $\Delta E$ = change in earnings and $\Delta PE$ = change in the PE ratio. A growth style focuses on the earnings growth, $\Delta E$. The value style focuses on the PE ratio and the dividend yield, while paying less attention to growth in earnings.

Attractive growth stocks show consistent earnings and growth of earnings. The price to earnings ratio of growth stocks is greater than that of the average stock. Typical growth stocks have a return on equity greater than 18%.

Value investors search for undervalued stocks by carefully examining the balance sheet and price to earnings ratio of a stock. Value stocks are corporate shares that sell

**Exhibit 5.1   Characteristics of Value and Growth Investing**

| Value Investing | Growth Investing |
|---|---|
| ■ Low or modest growth of earnings | ■ High growth of earnings |
| ■ Low PE ratio | ■ High PE ratio |
| ■ Low price to cash flow | ■ High price to cash flow |
| ■ Low price-to-book ratio | ■ High price-to-book ratio |
| ■ Low price-to-sales ratio | ■ Volatile prices |
| ■ Less volatile prices | ■ Consistent or accelerating growth rate |
| ■ Commodities or products in competitive markets | ■ Companies with dominant products and market share |
| ■ Debt-to-capital ratio less than 50% | |

for less than the company's intrinsic or true worth. Another important measure of value is a low price-to-sales ratio. The best growth stock always sells for a higher price-to-earnings ratio (PE) compared to a value stock. Nevertheless, slow-growing value stocks can do well over the long run.

Historically, growth investing has done better in times of low inflation and low interest rates. In times of higher inflation, greater than 4%, and climbing interest rates, value investing may be preferable. Thus, if we plan to hold our stocks over the long run, we need to combine the best of style of investing in our method of investing.

The characteristics of value and growth investing are shown in Exhibit 5.1. There are periods of history which have favored one investment style over the other. Therefore, we seek to utilize both strategies within our overall portfolio. Stocks that show stable, persistent earnings growth at a rate of 12% to 20% are attractive candidates for a part of the growth portion of our portfolio since these growth rates may be sustainable for a decade or more. A limited number of companies with growth rates greater than 20% may also sustain their growth rate for many years. However, they may be risky since any stumble in the earnings growth will result in a large drop in their PE ratio and price. An example of a high growth stock is Cisco Systems, which provided an average compound return of 66.8% for the 5-year period 1993–98. Cisco had a PE ratio of 64 in early 1999, while the expected growth rate for '99–'02 was 28%.

On the other hand if we can identify companies with an earnings growth rate that is slowly increasing, this growth stock may be undervalued. So growth is good when it is consistent, but even better when it is accelerating.

A size-versus-style chart is shown in Exhibit 5.2. Most mutual funds select a size (capitalization) and a style and operate in one of the quadrants of the chart. We will operate across the total chart. The large-cap stocks move together and the mild-cap stocks move together. However, the large-and mid-cap stocks often do not totally cor-

**Exhibit 5.2   Characteristics of Value and Growth Investing**

| Mid cap—Growth | Large cap—Growth | Growth | ⎫ |
|---|---|---|---|
| Mid cap—Value | Large cap—Value | Value | ⎬ Style |
| Mid | Large | | ⎭ |

**Size**-Capitalization

relate in response to market changes. Similarly with stocks selected for value and other stocks selected for growth we can achieve some diversification benefits. With diversification across all quadrants of the chart we can expect some benefits of noncorrelation between quadrants [O'Shaughnessy, 1997]. Whatever mix of style and size we select, our goal is to outperform the market return with a similar or lower risk.

Thus, we will use this simple method of selecting a diversified set of stocks. We use a value strategy to select a set of stocks from the mid-cap and the large-cap stocks. Then, we use the growth method to select high growth stocks from the mid-cap and large-cap stocks. In this book we use the price to earnings and the price to cash flow multipliers as the valuation multiplier since it is readily available in the Value Line Survey or from other sources.

## Sectors and Industries

The Value Line Investment Survey covers 96 industries such as machinery, electronics, and publishing. Standard and Poor divides the S&P 500 into 11 sectors as shown in Exhibit 5.3. These sectors incorporate all 96 of the Value Line industries. The S&P 500 is a portfolio put together by Standard and Poor and incorporates leadership companies from leadership sectors. Note the change in the weighting of sectors from 1964 to 1999. In 1964, utilities was the largest weighted sector while technology became the largest weighted sector in 1999. The S&P 500 portfolio has become more diversified, less cyclical and more global and it now possesses a greater degree of sustainable competitive advantage than ever before.

Different industries have different profitability levels and these differences persist over time. For example, it has been more profitable to make new pharmaceuticals than generic, over-the-counter drugs. Of course, we can still find a successful firm in less attractive industries due to that firm's management or relative advantage.

**Exhibit 5.3   The Market Capitalization Weight (%) of the Sectors of the S&P 500 for 1964 and 1999**

| | Market Cap Weight (%) | | 1999–2002 | |
| --- | --- | --- | --- | --- |
| Sector | 1964 | 1999 | Expected Growth Rate Earnings (%) | Expected Return (%) |
| Finance | 0.0 | 14.8 | 11 | 14 |
| Health | 2.3 | 11.8 | 13 | 16 |
| Consumer Nondurables | 10.0 | 9.9 | 12 | 15 |
| Consumer Services | 6.3 | 12.6 | 13 | 15 |
| Consumer Durables | 11.3 | 1.9 | 8 | 13 |
| Energy | 17.8 | 5.8 | 8 | 12 |
| Transportation | 2.6 | 1.0 | 8 | 11 |
| Technology | 5.5 | 19.0 | 17 | 21 |
| Basic Materials | 16.5 | 3.7 | 8 | 11 |
| Capital Goods | 8.5 | 8.0 | 11 | 13 |
| Utilities | 19.2 | 11.5 | 7 | 11 |
| **Grand Total:** | **100.0%** | **100.0%** | | |

**Table 5.1    Sectors**

**Attractive Sectors:**

| | |
|---|---|
| Technology: | Computers, semiconductors, communications, software, biotech |
| Publishing: | Magazines, newspapers, advertising |
| Financial: | Banks, securities, insurance |
| Health: | Drugs, instruments, equipment |
| Industrial: | Machinery, components, services |
| Food: | Processors, distributors, retailers, beverages |
| Transportation: | Airlines, trucks, railroads, marine, air freight |
| Retail Stores: | Discount, department |
| Autos: | Manufacturers, parts, tools, tires |
| Office Equipment: | Copiers, furniture |
| Home: | Furniture, tools, services, toiletries, cosmetics |
| Leisure: | Movies, hotels, toys |

**Less Attractive Sectors:**

| | |
|---|---|
| Basic Materials: | Metals, chemicals, mining, paper, steel |
| Energy: | Coal, oil, pipelines, generating plants |
| Utilities: | Electric, gas, water |
| Real Estate: | Real estate investment trusts |

Attractive sectors are listed in Table 5.1. Within each sector are listed some of the industries comprising that sector. Perhaps the most attractive sector is the technology sector. Health care and financial stocks are also attractive. A classic contrarian sector is energy, where we might expect oil and gas to firm at a base price and rise again to more normal price levels by 2000 or 2001.

## Asset Allocation

Asset allocation methods seek to exploit differences in the way assets fluctuate in sectors, industries, large- and mid-cap stocks, and between value, growth, and contrarian styles. It is the fluctuation or variance of the investor's total portfolio that represents the risk of the portfolio. The selection of types of stocks and their normalized weights as a proportion of the total portfolio leads to the total portfolio return and associated risk. We strive to select industries, sectors and stocks that do not rise and fall together (risk) and yield the expected return commensurate with that risk. Asset allocation utilizes expectations based on historical trends and is limited in accuracy by drift over time in the trends for sectors and industries.

The potential investment classifications we can consider are listed in Table 5.2. We can select stocks by 1) capitalization, 2) style, 3) risk and return, 4) sector and industry, and 5) geography. We can diversity our portfolio assets over all these classifications in order to satisfy our desired long-run return with an expected probability of achieving it and a volatility of the value of the portfolio that we can live with.

In order to build the final portfolio we follow the six steps outlined in Table 5.3. We first decide on the expected return and the acceptable risk. Clearly, we must determine an expected return consistent with the acceptable risk. It is unreasonable to expect an annual return of 30% over a 10-year period with a low risk level (and associated

**Table 5.2   Potential Investment Classifications for a Portfolio**

| | |
|---|---|
| ■ Attractive sectors: | Large-cap, Mid-cap, Small-cap |
| ■ Investment style: | Value, growth |
| ■ Risk and return subportfolios: | High growth leaders, high risk |
| |   Moderate growth blue chips, moderate risk |
| |   Low steady growth—low risk |
| |   Contrarian, out of favor, turnaround companies |
| |   Cash and Short-Term Bonds |
| ■ Sector and industry: | See Table 5.1 |
| ■ Geographic: | U.S., Europe, and Asia: Emerging Markets |

**Table 5.3   Asset Allocation Steps to Build a Portfolio**

1. Determine the expected return and acceptable risk of the portfolio
2. Select asset classes to include in the portfolio
3. Consider long-term returns and risk of each class and estimate covariance between classes
4. Determine what proportion (weight) of the portfolio should be assigned to each class
5. Select individual stocks within each class
6. As circumstances change, adjust the weighting of each class

volatility of portfolio value). The second step is to select the investment classes we will include. For example, for our purposes, we will not include stocks from emerging market countries. The third step is to estimate the long-term returns expected from each investment class and the expected covariances. The fourth step is to determine the proportion of the portfolio to be assumed by each class. The fifth step is to select stocks within each class. The sixth step is to adjust the portfolio over time so as to return the prepared portfolio balance. The asset allocation steps of Table 5.3 can be performed using asset allocation software. However, we will illustrate a straightforward manual process.

## Building a Diversified Stock Portfolio Using Four Subportfolios

By reducing the number of investment classes we can use an intuitively attractive allocation process. Let us assume that we only will use large-cap and mid-cap stocks. Secondly, we will use a mixed value and growth style. Thirdly, we will diversify among attractive industries. Finally, we will primarily invest in U.S. companies, while reserving a small portion of our portfolio for large foreign blue chip companies. With these assumptions we create our total portfolio using the four subportfolios listed in Table 5.4. These four subportfolios are shown on the risk-versus-return chart of Figure 5.1. The four subportfolios are called by a name characteristic of their attributes and are labeled: Gazelles, Hares, Tortoises, and Contras. The total stock portfolio consists of these four subportfolios. In addition, we will hold a subportfolio consisting of cash and short-term bonds. Typically, the cash portion of your folio should consist of cash necessary for the next six to 12 months. We will assume this cash is available as necessary and now proceed to allocate the remaining four subportfolios to add together as your stock portfolio. The four subportfolios are listed in Table 5.4 with their attributes and sample companies.

**Table 5.4    Four Subportfolios Used to Assemble a Total Portfolio**

<u>Gazelles</u>
High Growth, Growth Rate of Earnings > 16%, PE > 20
Expected High Annual Return > 20%
Examples: Microsoft, Starbucks, and EMC Corp.
Selected using growth criteria

<u>Hares</u>
Moderate Growth, Growth Rate of Earnings between 10% and 15%
PE typically between 10 and 20
Market Leaders, Blue Chips
Examples: General Electric, MacDonalds, Walgreen
Selected using growth with value criteria

<u>Tortoises</u>
Low Growth, Growth Rate of Earnings between 4% and 9%
PE typically between 8 and 10
Utilities, REITs, mature companies
Examples: New Plan Excel, Nicor and Duke Energy
Selected using value with growth criteria

<u>Contras</u>
Contrarian, out of favor, turnaround companies
Low price to earnings, low price to book
Companies formerly strong, but earnings have declined significantly
Examples: K-Mart, Eastman Kodak, Halliburton, and Nike
Selected using value criteria

The Gazelle Portfolio, with Index 1, includes high-growth companies such as Microsoft and EMC Corp. The moderate-growth portfolio Hares, with Index 2, include companies like General Electric and Gillette. The low-growth portfolio Tortoises, Index 3, include utilities, REITs and mature companies such as Duke Energy, New Plan Excel, and Nicor. The Contra portfolio, Index 4, includes out of favor and turnaround companies such as Nike and Eastman Kodak.

The Total Stock portfolio is then assembled from the four subportfolios with a weighting for each portfolio is selected. Thus, $w_1$ = the portion of the portfolio allocated to Gazelles. Similarly, $w_2$ is the weight for the Hare subportfolio, $w_3$ is for the Tortoises and $w_4$ is the weight for the Contra subportfolio.

The estimated standard deviation (risk) and return for each of the subportfolios is listed in Table 5.5. For ease of calculation, we first build a subportfolio, with Index m, consisting of a combination of the Gazelles and Hares as shown in Table 5.6. We select $w_1 = 0.3$ and $w_2 = 0.7$ as a reasonable lower-risk combination. Then, the return of this subportfolio is

$$R_m = w_1R_1 + w_2R_2 = 0.3(20) + 0.7(14) = 15.8\%.$$

The subportfolio n consists of the Tortoise and the Contra portfolios combined with weights $w_3$ and $w_4$. The return for subportfolio n is calculated using $w_3 = 0.7$ and $w_4 = 0.3$. Then, $R_n$ can be calculated as

$$R_n = w_3R_3 + w_4R_4 = 0.7(10) + 0.3(12) = 10.6\%.$$

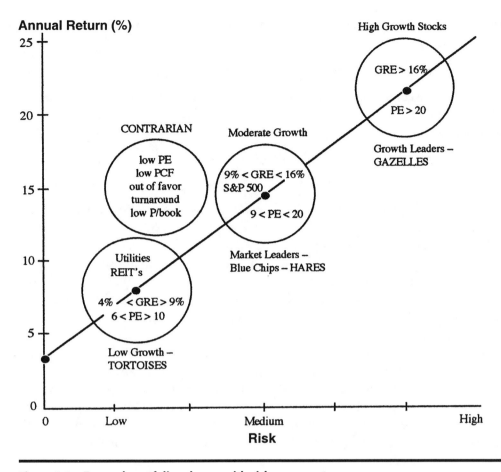

**Figure 5.1** **Four subportfolios shown with risk versus return**

---

**Table 5.5** **The Expected Risk and Return for the Four Subportfolios for the next five-year period**

| Subportfolio | | | | |
| --- | --- | --- | --- | --- |
| Name | Gazelles | Hares | Tortoises | Contra |
| Number | 1 | 2 | 3 | 4 |
| Expected Return | 20% | 14% | 10% | 12% |
| Expected Risk, Standard Deviation | 24% | 16% | 8% | 10% |

---

**Table 5.6** **Correlation, $\rho_{ij}$, between Subportfolios**

| Portfolio | Name | Gazelle | Hare | Tortoise | Contra | | |
| --- | --- | --- | --- | --- | --- | --- | --- |
| | Number | 1 | 2 | 3 | 4 | Name | Number |
| | | 1 | 0.8 | 0.5 | 0.5 | Gazelle | 1 |
| | | | 1 | 0.5 | 0.5 | Hare | 2 |
| | | | | 1 | 0.8 | Tortoise | 3 |
| | | | | | 1 | Contra | 4 |

The risk, standard deviation of return, is calculated from the equation for subportfolio m as [Sharpe, 1995]

$$\sigma_m^2 = (w_1\sigma_1)^2 + (w_2\sigma_2)^2 + 2\,w_1w_2(\rho_{12}\sigma_1\sigma_2)^2 \qquad (5\text{--}1)$$

where $\rho_{12}$ is the **correlation** between subportfolios 1 and 2. The correlations between all four subportfolios are given in Table 5.6. Note that $\rho_{12} = 0.8$. Therefore, the calculation for $\sigma_m^2$ is:

$$\sigma_m^2 = (0.3 \times 24)^2 + (0.7 \times 16)^2 + 2(0.7 \times 0.3)(0.8 \times 24 \times 16)$$
$$= 51.84 + 125.44 + 129.02 = 306.30.$$

Therefore, $\sigma_m = 17.50$. Using a similar calculation for subportfolio n which consists of a combination of Tortoises and Contra with a correlation $\rho_{34} = 0.8$, we determine the return and risk for subportfolio n as shown in Table 5.7.

We now combine subportfolios m and n into a total portfolio using the weights $w_m$ and $w_n$ as shown in Table 5.8. Examining Table 5.6 we deduce that the correlation $\rho_{mn} = 0.5$. The resulting return and risk for various sets of weights is given in Table 5.8. Since the correlation coefficients $\rho_{12}$, $\rho_{34}$, and $\rho_{mn}$ are all less than one, the risk is reduced from the situation where $\rho = 1$.

We plot the risk versus return for the total portfolio in Figure 5.2. The curve shown in the Figure 5.2, called the efficient frontier, represents the total portfolio for the various weights listed in Table 5.8. Note that the curve for the total portfolio lies above the return for the Hare portfolio. Thus, we gain additional return for the same risk or have lower risk for the same return. When $\sigma = 16\%$ we obtain an improved return of 15.2% compared to the Hare return of 14%. Diversification with the 4 subportfolios provides higher returns with less risk.

**Table 5.7    The Expected Risk and Return for the Four Subportfolios for the Next Five-Year Period**

|  | Subportfolio m | | Subportfolio n | |
|---|---|---|---|---|
| Name | Gazelles | Hares | Tortoises | Contra |
| Number | 1 | 2 | 3 | 4 |
| Weighting | 0.3 | 0.7 | 0.7 | 0.3 |
| Subportfolio Return | 15.8% | | 10.6% | |
| Subportfolio σ | 17.5% | | 8.2% | |

**Table 5.8    The Risk and Return for the Total Portfolio with Various Weightings**

| Risk | Weights | | Final Portfolio Return (%) | Final Portfolio σ (%) |
|---|---|---|---|---|
|  | $w_m$ | $w_n$ | | |
| Conservative | 0 | 1 | 10.6 | 8.2 |
| Mid-Conservative | 0.3 | 0.7 | 12.2 | 9.5 |
| Moderate | 0.5 | 0.5 | 13.2 | 11.4 |
| Mid-Aggressive | 0.7 | 0.3 | 14.2 | 13.6 |
| Near-Aggressive | 0.8 | 0.2 | 14.8 | 14.9 |
| Aggressive | 1 | 0 | 15.8 | 17.5 |

## Selecting Your Best Total Stock Portfolio

The final step is to select the best portfolio from the possibilities on the efficient frontier shown in Figure 5.2. For the aggressive investor with a time horizon longer than 15 years, the portfolio with $w_m = 1$ and $w_n = 0$ may be appropriate. However, this aggressive investor should be prepared to stick with his holdings through a one- or two-year decline of 40% or more at some time in the 15-year period.

The selection of the best portfolio on the efficient frontier is facilitated by selecting the risk, reward and holding period profile for your portfolio. Table 5.9 lists a set of portfolios with an estimated risk and reward profile. For a holding period of 12 to 15 years with significant risk, one might select $w_m = 0.7$ and $w_n = 0.3$. Then, the weighting for the Gazelle subportfolio is

$$w_1 = 0.3 \times w_m = 0.3 \times 0.7 = .21$$

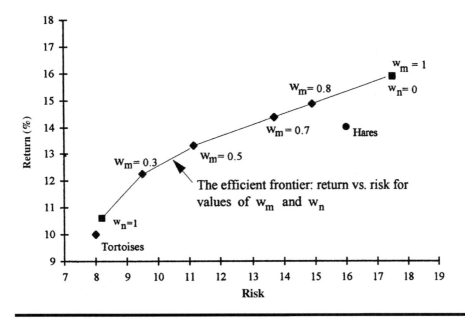

**Figure 5.2   The return of the total portfolio contrasted with the return for the Hare and Tortoise portfolio.**

**Table 5.9   The Set of Acceptable Portfolios**

| Weights | $w_m$ | 0.3 | 0.5 | 0.7 | 0.8 |
|---|---|---|---|---|---|
| | $w_n$ | 0.7 | 0.5 | 0.3 | 0.2 |
| Holding Period (Years) | | 5–8 | 8–12 | 12–15 | 15 or longer |
| Risk | | Low | Medium | Significant | High |
| Return | | 12% | 13% | 14% | 15% |
| Potential 1-Year Decline | | 10% | 15% | 20% | 25% |

and the weighting for the Hare portfolio is

$$w_2 = 0.7 \times 0.7 = 0.49.$$

In a similar manner we calculate $w_3 = 0.21$ and $w_4 = 0.09$. This might prove to be our best portfolio for this holding period, the expected return and an acceptable risk.

Once we have chosen the appropriate mix of subportfolios, we then turn to the selection of stocks appropriate for each subportfolio. Remember, our plan is to select five to ten companies that will be most likely to produce above-average returns within each subportfolio. Thus, we expect to achieve above-market returns for our desired overall risk level. Our goal is to reduce the risk of our overall portfolio falling behind the market return while increasing its probability of outperforming it.

# 6 Valuation Measures

## This chapter covers

- The Valuation of a Company
- Buying a Business
- Buffett's Methods
- Return Expectations and Profit Margins
- The Aging Process—Four Phases
- Return on Equity
- The PE Ratio
- The PEG Ratio
- Return on Assets
- The Discounted Earnings Model

The selection for purchase of the common stock of a company can be based on three valuation methods:

- Intrinsic Value
- Relative Value
- Acquisition Value

The **intrinsic** (real) **value** of a firm is the net present value of its future cash flows. The **relative value** of a firm is determined by comparing it to similar companies' values. The **acquisition value** of a company is the worth of the company to an acquiring firm.

The acquisition valuation method suffers from the limited ability of an individual investor to secure all the information that a professional acquirer can find. Also, acquirers will ultimately be managing the acquired company or hold seats on the board of directors and will set the direction for the company. This method is not suitable for the individual and we will only consider the intrinsic value and relative value methods.

Fundamental analysis assumes that the intrinsic value of any common stock equals the present value of all cash flows that the owner expects to receive. The intrinsic value of a common stock is defined as its real or warranted value. Thus, the owner attempts to forecast the timing and size of these cash flows and then converts them to their present value by using an appropriate discount rate. On this basis, the analysis provides an intrinsic value which can be compared to the current stock price. If the stock price is less than the intrinsic value, then this company stock is a good candidate for purchase. If the price is below its intrinsic value, the tendency is for the market to recognize this mispricing and correct the price to the real value. The primary objective of the capital markets is to efficiently allocate capital to companies that produce the greatest wealth. Thus, the value of a company is directly related to its ability to create wealth and this ability can be discerned by evaluating the intrinsic value. However, the individual investor may be limited to less than full information and the company managers may not always maximize shareholder value.

A company's earning power defines its future prospects. For many companies, however, it is difficult to predict the earnings growth of the company. Therefore, many use the relative valuation method. This method records the valuation parameters such as PE ratio, price-to-book ratio and profit margin of the candidate company and the same parameters for one or two similar companies in the same industry. If the selected ratios are better than those of the comparable companies, the selected company is said to be undervalued. This method is often complicated by the lack of good comparables and the whole industry may be overvalued. This method uses past and current data and often is not sensitive to the dynamics of the market.

The valuation of a common stock is not an exact science. The dynamics of the market and new information are constantly received. For example, Hewlett-Packard announced a significant restructuring of the company on March 2, 1999, and the market value increased due to the perception of improved value in the future. The market perceived better future returns and reallocated capital to this higher potential firm.

## Buying a Business

An investor seeks to identify an excellent business in an attractive industry. A great company can be held for the long term and provide excellent return on your investment. If you were acquiring the total ownership of a business you would seek a business in an industry you could understand and with qualities that can be readily determined. Exhibit 6.1 lists the qualities of a company that can lead to a valuable investment.

Great companies exhibit many of these qualities. In the book, *The Living Company,* de Geus defines a living company as sensitive to its environment, cohesive, tolerant,

**Exhibit 6.1   The Qualities of an Outstanding Business**

- Clear vision and sense of purpose
- Strong sense of identity
- Investment for the long term
- Pursuit of short-term and long-term performance
- Commitment to a core ideology
- Creates business systems for the future
- Selects leadership from within
- Financial conservation
- Sensitive to the business environment and adaptable to change
- Tolerance of individual employee entrepreneurs
- Strong corporate culture
- High standards of employee performance

and conservatively financed [de Geus, 1997]. With a focus on intellectual capital and learning, the employee is key to success. The living organization possesses its own personality with a character and history. The company with the long-term view regards the optimization of the use of capital as a complement to the continual improvement of the skills and abilities of their people. The unwritten contract is that the individual will deliver skill, commitment, and care in exchange for the company striving to develop each person's potential to the maximum.

In an equally valuable book, *Built To Last,* the authors describe the visionary company which displays great resiliency and performance [Collins, 1994]. They show that the company and its vision are the fundamental creation. With an attendant culture and big goals the company persists in fulfilling its vision while adapting to challenges, finding pragmatic solutions, and behaving consistent with its core values.

An example of a living company built to last is General Electric. Founded in 1892, the company has consistently pursued new, valuable technology and innovation and responded to change. It has promoted from within and strives to enhance its people's skills and knowledge.

## Buffett's Methods

The investor, Warren Buffett, is widely followed as one of the most effective investors of modern times. Buffett seeks to identify excellent companies and buy and hold their stock. He seeks companies earning at least 15% on equity capital. He selects companies that have many of the qualities of Exhibit 6.2 [Hagstrom, 1994]. Using calculated intrinsic values, Buffett evaluates the value of attractive companies. Buffett purchased See's Candy Shops and Buffalo News using these criteria.

Buffett noted that there are only five ways for companies to increase their return on equity as listed in Exhibit 6.3. The last two ways (lower taxes and cheaper leverage) are usually beyond the control of the company. If the company significantly increases the leverage, it violates the need for financial safety. Increasing the ratio of sales to net property and plant, and the operating margin, are sound actions and many attractive companies have higher ratios. Buffett's goal is a return of 15% per year over a 15-year period. Examples of Buffett's companies are Gillette, Coca-Cola, and *The Washington Post.*

### Exhibit 6.2 Qualities of Companies for Buffett's Method

- Return on equity of 14% or greater
- Invest like you are buying the business
- Consumer monopolies selling products where there are few competitors and not regulated
- High profit margins; profits to sales ratio
- Good brand name
- Readily understandable business
- Consistent earnings history
- Low financial leverage; assets/equity and debt/capital
- Earnings growth on upward trend
- Able to adjust prices to inflation
- Favorable long-term prospects
- Important company in growing industry

**Exhibit 6.3   Ways to Increase Return on Equity**

- Increase ratios: sales/assets and sales/net plant & property
- Increase operating margin*
- Increase leverage
- Use cheaper leverage
- Pay lower taxes

*Operating Margin = earnings before depreciation, interest, and taxes/sales.

Buffett seeks companies with monopolylike qualities which many call "franchises." A single daily newspaper in a regional market is an example of a business franchise. Buffett uses "owner earnings" to calculate the value of a company which he defines as: net profits plus depreciation minus annual capital spending. If we exclude capital spending we have owner earnings = cash flow.

Another valuation method used by Buffett utilizes the yield on the long-term 10-year Treasury Bond, I, and next year's expected earnings, E, of the stock under consideration. Then, a first estimate of intrinsic value is equal to the ratio of E to I. Therefore,

$$\text{Intrinsic Value} = \frac{E}{I} = PE \times E.$$

Note that the calculated PE ratio is equal to 1/I. For example, consider AT&T in early 1999 with an expected E = $4 for 1999. At that time R = 0.05 (5%) and therefore,

$$\text{Intrinsic Value} = \frac{E}{I} = \$4/.05 = \$80.$$

The actual price of AT&T in early 1999 was $81. Thus, AT&T was fairly valued at that time.

As yields on Bonds decline, investors find it easier to justify higher PE multiples since PE = 1/I is an approximation to reality. The two driving forces for stock prices are earnings growth and interest rates. The best conditions occur when earnings are growing steadily and interest rates declining steadily (these occurred in 1997-99).

## Return Expectations and Profit Margins

The investor seeks solid returns based on reasonable performance expectations for a company. The responsibility of the investor is to understand, evaluate, and anticipate changes. Obviously, the investor seeks companies with positive changes in their performance. In the end, earnings expectations drive stock prices, along with interest rates driving price multiples. The market discounts expected changes in corporate profitability before those changes occur. The market readily discounts changes in earnings (earnings growth) as well as changes in earnings growth (earnings accelerations). The secret of great investing is an ability to spot and seize the upside surprise in earnings. This difficult task is best accomplished in industries you understand and for which you have access to broad discussion of trends and new activity. One possibility is to keep a file on the top companies in the three or four industries you follow.

Common measures of profitability are profit margin and operating margin. These are both provided in the Value Line Survey or the S&P reports. The profit margin is defined as the ratio of profits to sales for a given period, normally a fiscal or calendar year. Thus, profit margin, PM, is

$$PM = E/S$$

where E = earnings for the year and S = sales or revenues for the year. Operating Margin (OM) is operating earnings (before deduction of depreciation, interest and income tax) as a percentage of sales or revenues. Thus, OM is

$$OM = OE/S$$

where OE are operating earnings.

Table 6.1 shows the five-year ('93–'98) average profit margin and operating margin versus the annual average compound return for companies in three attractive industries. They are ranked in order of annual return for the five-year period. Clearly, General Electric earned a strong return by exhibiting a high profit margin and a high operating margin. Costco exhibited a strong return with margins below Nordstrom. Finally, Microsoft doubled the return of Computer Associates while showing margins similar to those of Computer Associates.

We can conclude from Table 6.1 that companies with strong margins will do well compared to companies with weak margins as exhibited by comparing WalMart to K-Mart. Since we are interested in trends in margins, let us look at the profit margin for selected companies during the five years '93–'98 as well as the expected profit margin for 1999 (see Table 6.2).

General Electric exhibits a higher and growing profit margin while Johnson Controls exhibits a lower and flat profit margin. Thus, General Electric possesses a more attractive record and should be valued more highly over the period.

**Table 6.1. The Five-Year ('93–'98) Average Profit Margin and Operating Margin versus Annual Return for Companies in Selected Industries**

| Industry, Company Name | Profit Margin (%) | Operating Margin (%) | Return (%) |
|---|---|---|---|
| **Electrical Equipment** | | | |
| General Electric | 16.1 | 18.4 | 34.2 |
| Johnson Controls | 2.4 | 8.6 | 20.0 |
| Honeywell | 5.4 | 13.2 | 19.5 |
| Emerson Electric | 9.1 | 19.4 | 17.6 |
| **Retail Stores** | | | |
| Costco | 1.4 | 3.6 | 30.3 |
| WalMart | 3.1 | 5.9 | 27.6 |
| Nordstrom | 4.1 | 8.3 | 17.2 |
| J. C. Penney | 3.6 | 8.0 | 1.5 |
| K-Mart | 0.7 | 3.9 | −4.6 |
| **Software** | | | |
| Microsoft | 27.8 | 45.4 | 68.9 |
| BMC Software | 30.8 | 46.7 | 49.3 |
| Computer Associates | 23.4 | 49.0 | 29.5 |
| Auto Data Processing | 13.0 | 24.7 | 24.9 |
| Autodesk | 13.5 | 25.7 | 14.5 |

**Table 6.2   Profit Margin for '93–'99 ('99 estimate)**

| Name | '94 | '95 | '96 | '97 | '98 | '99 (Est.) |
|------|------|------|------|------|------|------|
| General Electric | 14.9 | 15.3 | 15.8 | 16.8 | 17.7 | 18.5 |
| Johnson Controls | 2.4 | 2.4 | 2.3 | 2.4 | 2.4 | 2.4 |
| Costco | 1.2 | 1.2 | 1.3 | 1.6 | 1.9 | 1.8 |
| J. C. Penney | 5.2 | 4.1 | 3.5 | 2.8 | 2.2 | 2.5 |
| K-Mart | 0.3 | −1.5 | 0.8 | 0.9 | 1.6 | 1.8 |
| Microsoft | 26.0 | 24.5 | 25.1 | 30.4 | 33.0 | 35.4 |
| Autodesk | 16.1 | 16.4 | 9.0 | 11.7 | 14.2 | 15.6 |

Costco exhibits a growing profit margin compared to J. C. Penney which shows a declining profit margin and K mart which experienced a deficit in 1995 and is only slowly recovering in the period '98–'99.

Finally, Microsoft shows an increasing profit margin while Autodesk shows a decline in '96–'97.

We seek companies with solid margins compared to the margins of their competitors in their industries. Similarly, we are looking for growing margins and evidence of increasing profitability. It is no surprise that Microsoft and General Electric are highly valued performers in the market.

## The Aging Process—Four Phases

Companies exhibit a life cycle from birth to senior citizen. The four phases of a company's life are shown in Table 6.3. Very agile companies can remain in the mature phase for many decades by continuing to adapt to changes in the competitive marketplace.

It is helpful to identify the phase that a company has grown into. For example, IBM is a mature, large company. We look for consistent earnings and returns from IBM and other companies that have reached the mature phase. As a company matures its profit margins increase and then level off at a consistent level.

The early phase companies are interested in building revenues and market share. Typically, these companies spend to build their products and business and exhibit losses. A good example is Amazon.com that is building capabilities and revenues but is yet to show a profit as of 1999. However, the company has grown revenues from $15 million in 1996 to over $1 billion in 1999.

Many well-known companies are in the mature phase such as IBM and General Electric. The earnings of these companies grow at about 14% per year and the profit margins are typically 10% or greater. Note that the profit margin of General Electric is about 16%.

The slow growth or declining growth company is a defensive holding for the portfolio and normally fits into the Tortoise category. Examples are utilities and steel companies.

**Table 6.3   The Four Phases of a Company**

| Phase | Early | High Growth | Maturity | Slow or Declining Growth |
|---|---|---|---|---|
| Sales | Less than $100 million | $100 million to $10 billion | Greater than $10 billion | Greater than $10 billion |
| Sales Growth | Greater than 30% | Greater than 20% | Greater than 10% | Less than 10% |
| Earnings Growth | Losses or low profits | Greater than 20% | Greater than 10% | Less than 10% |
| Profit Margin | None or low | High Greater than 15% | Moderate 10% to 15% | Low Less than 10% |
| Consistency of Profits and Return | None or low | Accelerating profitability and returns | Potential for consistent growth of profits and consistent returns | Declining profitability |
| Risk of Negative Occurences | High | Significant | Moderate | Significant |
| Examples of Companies in Selected Industries | Uniphase Optical Cable Amazon.com Human Genome Science | Microsoft Bed Bath & Beyond Dell Computer Gap | General Electric DuPont New York Times IBM | Consolidated Edison K-Mart Kodak |
| Examples of Industries in each Phase | Fiber Optics Biotechnology Internet | Computer Software Health Services | Electrical Equipment Chemicals Newspapers | Steel Textiles Electric Utilities |
| Classification of Companies | Gazelles | Gazelles and Hares | Hares and Tortoises | Tortoises and Contras |

# Return on Equity

A key measure of profitability is return on equity (ROE) which is

$$\text{ROE} = \text{E/S} \times \text{S/A} \times \text{A/Eq}$$

where E = earnings, S = sales, A = total assets, and Eq = equity. The first factor is profit margin, E/S. The second factor is asset turnover, S/A, which is the ratio of sales to assets. The final factor is the financial leverage represented by A/Eq. The company that generates high ROE through high return on sales and asset turnover is in a good position. A company that uses high financial leverage is at risk from significant indebtedness. The average financial leverage ratio is 2.6 for the S&P 500 while the average financial company, such as banks, have a ratio of 12. Utilities have ratio of about 3.0.

The return on equity calculation using the three factors is illustrated in Table 6.4. Four companies are listed in sequence of increasing profit margin. Note that Sysco has a low profit margin but overcomes that limitation by a high asset turnover. Microsoft achieves a high ROE through a very high profit margin. Microsoft has a low financial leverage. A financial leverage greater than 3.5 may expose a company to excessive risks.

The return on equity is an important measure of performance and will correlate with the total return of the stock. Table 6.5 shows the average ROE and average annual

**Table 6.4    Return on Equity for Four Companies for 1999**

| Name | Profit Margin × E/S × | Asset Turnover × S/A × | Financial Leverage = A/Eq = | ROE ROE |
|------|------------------------|--------------------------|------------------------------|---------|
| Sysco | 350/17,300 | 17,300/3,780 | 3,780/1,585 | |
| | 2.02% | 4.58 | 2.38 | 22.0% |
| Miller, Herman | 140/1,865 | 1,865/784 | 784/305 | |
| | 7.51% | 2.38 | 2.57 | 45.9% |
| Clorox | 345/2,950 | 2,950/3,030 | 3,030/1,250 | |
| | 11.69% | 0.97 | 2.42 | 27.4% |
| Microsoft | 6,200/17,500 | 17,500/14,387 | 14,387/21,465 | |
| | 35.43% | 1.22 | 0.67 | 29.0% |

Source:   Value Line and company reports

**Table 6.5    Average Return on Equity and Average Annual Return for Selected Stocks for the Five-Year Period 1993–98**

| Company | Average Return on Equity (%) | Average Annual Return (%) | Direction of Change in ROE |
|---------|------------------------------|---------------------------|----------------------------|
| Atlantic Richfield | 16.2 | 9.3 | ↘ |
| Auto Data Processing | 19.1 | 24.9 | → |
| Aflac | 14.9 | 37.3 | ↘ |
| Best Foods | 39.7 | 21.7 | ↗ |
| Dayton Hudson | 14.3 | 39.7 | ↗ |
| Emerson Electric | 19.5 | 17.6 | → |
| General Electric | 23.5 | 34.2 | → |
| Gap | 33.4 | 46.3 | ↗ |
| Harley-Davidson | 22.0 | 34.6 | → |
| Halliburton | 13.8 | 16.6 | ↗ |
| Intel | 29.8 | 50.6 | → |
| Long's Drug | 10.0 | 21.0 | → |
| Mattel | 25.7 | 11.6 | ↘ |
| Microsoft | 29.3 | 68.9 | → |
| Miller, Herman | 25.1 | 30.3 | ↗ |
| Minnesota Power | 9.7 | 13.5 | ↗ |
| Nordstrom | 13.1 | 17.2 | ↘ ↗ |
| Nicor | 15.9 | 13.3 | → |
| Oracle | 35.3 | 38.3 | → |
| J. C. Penney | 13.3 | 1.5 | ↘ |
| Pfizer | 28.9 | 51.2 | → |
| Procter & Gamble | 27.0 | 28.6 | → |
| Schwab | 26.4 | 64.7 | → |
| Southwest Airlines | 15.2 | 6.6 | ↘ ↗ |
| Walgreen | 18.1 | 43.6 | → |
| Wal-Mart | 19.6 | 27.6 | ↗ |
| Xerox | 24.0 | 34.6 | ↗ |

return of selected stocks for the five-year period 1993–98. We also display the direction of change of ROE over that five-year period (up, down, or flat). This data is used to create Figure 6.1. There is a correlation between ROE and annual return over the five-year period as shown in Figure 6.1. An approximation of this relationship is

$$\text{Percent Return} = 1.8 \text{ ROE} - 7.1.$$

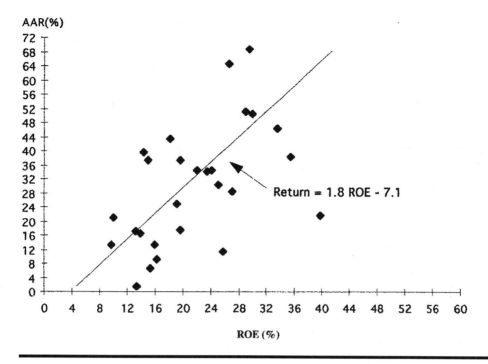

**Figure 6.1    Average Return on Equity versus Average Annual Return for the Five-Year Period 1993–1998**

Of course, other factors do influence the return. Companies with ROEs that decline over the five-year period, such as Mattel, will exhibit lower return than those companies, such as Dayton Hudson, that exhibited increasing ROEs over the five-year period starting in 1993. A record of the increasing ROE for Xerox for the period 1988–99 is shown in Table 6.6. Note the increase in ROE after 1993.

## The PE Ratio

The PE ratio is a measure of the expected return on a stock. The PE ratio is

$$PE = Price/Earnings = P/E.$$

A PE ratio uses the current price, P, and the earnings, E, over a 12-month period. The 12-month period can be the past 12 months or the next calendar year. *The Wall Street Journal* and other newspapers use the most recent four reported quarters as the

**Table 6.6    Return on Equity for 1988–1999 and Average Annual Return for Xerox**

| Year | '88 | '89 | '90 | '91 | '92 | '93 | '94 | '95 | '96 | '97 | '98 | '99est. |
|------|-----|-----|-----|-----|-----|-----|-----|-----|-----|-----|-----|---------|
| ROE (%) | 11.2 | 13.2 | 9.7 | 10.1 | 11.4 | 11.5 | 15.9 | 23.2 | 23.7 | 25.5 | 32.0 | 31.0 |
| **Average Annual Total Return (%):** | | | | | | 62.1 | | 39.8 | | 34.6 | | 24.1 |
| Period (Years): | | | | | | 1 | | 3 | | 5 | | 10 |

Periods ended: 12/31/98

one-year period. Value Line uses the actual earnings for the past six months and the estimated earnings for the next six months. Thus, the PE reported by Value Line will normally be less than the PE reported by newspapers.

The PE reflects how investors think a firm's future growth appropriately values the stock. The earnings, E, may be affected by one-time charges and other accounting factors and may not be the best indicator of value. The PE ratio is a statistic that incorporates the growth and risk aspects in one multiplier, PE. The PE ratio will change with changes in the perception of a company's characteristics.

We can look at the PE as a multiplier based on expectations of long-term growth of earnings, g, and the expected or required return, R. Then, one useful formula is

$$PE = \frac{1}{R - g}$$

where R and g are given in decimal form. Therefore, if we have a stock of General Electric we might have an expectation of earnings growth of 15% per year and a return expectation of 18% per year. Therefore, g = 0.15 and R = 0.18. Then, the associated PE ratio is

$$PE = \frac{1}{0.18 - 0.15} = \frac{1}{0.03} = 33.$$

As expectations change, the PE ratio changes. If General Electric slows down in growth rate to 13% and our expectation for return remains at 18% we have a drop in PE to

$$PE = \frac{1}{0.18 - 0.13} = \frac{1}{0.05} = 20.$$

This drop in PE is exactly what occurs when any company announces a slowdown in their earnings growth.

The PE ratio of the S&P 500 has had a mean of 14.5 over the past 30 years with a high in 1999 of 28 and a low of 8 in 1974 and again in 1979. The market in 1974 was forecasting a growth in earnings of 5.0% and an expected return of 13% (the corporate bond rate) so that

$$PE = \frac{1}{0.13 - 0.05} = \frac{1}{0.125} = 8.$$

On the other hand, the market in 1999 could be said to be forecasting an expected return of 15% and a growth rate of earnings of 11.4% so that

$$PE = \frac{1}{0.15 - 0.114} = \frac{1}{0.036} = 28.$$

However, the risk of not attaining the growth rate is reflected in an additional analysis when the growth actually achieved is g = 0.09 instead of the predicted g = 0.114. Then, the PE may drop to

$$PE = \frac{1}{R - g} = \frac{1}{0.15 - 0.09} = \frac{1}{0.06} = 16.7.$$

When the expected earnings growth rate drops by 21% the PE may drop by 40% as shown in this case. The PE ratio is sensitive to changes in the perceived growth rate of earnings, g.

We can expect this sensitivity to be useful if we choose a stock with a low PE ratio due to low growth rate expectations. If the growth rate picks up, then the PE ratio quickly reacts and prices quickly turn up. This fact is useful in the Contra portfolio where we have out of favor companies with low expectations for earnings.

Many companies in our Gazelle portfolio have high current PEs. Are these high current PEs justified? Let's define the price/earnings ratio based on the next year's anticipated earnings as the Forward PE (FPE). The anticipated growth rate in earnings is g and we assume the current earnings are E. Thus, next year's earnings are expected to be the forward earnings, FE, and then,

$$FE = (1 + g)E$$

and therefore

$$FPE = \frac{\text{Current Price}}{FE} = \frac{PE \times E}{(1 + g)E} = \frac{PE}{(1 + g)}.$$

For a high growth rate, say, 40%, g = 0.4 and

$$FPE = \frac{PE}{1.4} = 0.71\ PE.$$

If the current PE is 38 for a stock with g = 0.40, then,

$$FPE = 0.71\ PE = 0.71(38) = 27.0.$$

Therefore, the high current PE may be justified on the basis of the high growth rate of earnings.

For example, consider IBM with a current PE in March 1999 equal to 26. If we anticipate a growth rate of 30% we have g = 0.3 and then FPE = 20. In the early years of Microsoft after it became a public stock (1986) it had a growth of 100% (g = 1.0). Then,

$$FPE = \frac{PE}{(1 + g)} = \frac{PE}{2} = \frac{50}{2} = 25.$$

Looking at the reasonable FPE and the good earnings growth, an investor should have found it attractive to invest at that time.

## The PEG Ratio

Many investors use the PE to G ratio to determine fair value. For this ratio we use G as the growth rate of earnings given in percent. Thus, if a company has a PE = 24 and G = 18%, then,

$$PEG = 24/18 = 1.33.$$

At a PEG below 1.0, a stock may be seen as trading at a discount to its growth rate and thus may be a better bargain. The PEG ratio only has a reasonable usefulness for higher growth stocks. Obviously, for a typical utility with G = 5% and PE = 10, we have

$$PEG = 10/5 = 2.$$

Does this PEG = 2 disqualify this utility stock? Probably not. The greatest usefulness for the PEG measure is for companies with G greater than 9% and when stocks are compared within a single industry. A utility with PE = 8 and G = 8% so that PEG = 1 may be attractive. On the other hand, a company with PE = 40, G = 35%, and PEG = 1.14 may be fairly valued or slightly overvalued. Investors unwilling to buy stocks with a PEG greater than 1 may miss many excellent growth stocks.

Table 6.7 tabulates the average PE ratio and the expected earnings growth for 16 stocks using 1995 Value Line data. We then calculated the PEG ratio for each stock. Finally, we recorded the actual average annual return achieved by each company. If you created this table in 1995 you could have selected the stocks with a PEG of 1.10 or below and achieved an excellent return on these stocks. The actual average annual return is plotted versus the PEG ratio for all 16 stocks in Figure 6.2. The regression line for these data is approximately

$$Return = 106.9\text{-}62.9 \, PEG.$$

This regression line is also shown in Figure 6.2. We note that the five-year period 1993–98 was a period of abnormally high returns. Perhaps a more normal relation would be

$$Return = 45 - 30 \, PEG \tag{6.1}$$

as also shown in Figure 6.2.

**Table 6.7   The PEG Ratio and the Average Annual Return for the Period 1995–98**

| Name | Average 1995 PE | 1995–1998 Expected Earnings Growth (%) | 1995 PEG | 1995–1998 Average Annual Return (%) |
|---|---|---|---|---|
| Cisco | 20.7 | 50.0 | 0.41 | 77.5 |
| Selectron | 17.9 | 35.0 | 0.51 | 61.5 |
| Tellabs | 28.6 | 50.0 | 0.57 | 54.8 |
| Danaker | 17.2 | 29.0 | 0.59 | 51.0 |
| Microsoft | 28.2 | 40.0 | 0.70 | 84.9 |
| Gap | 15.0 | 19.0 | 0.78 | 83.1 |
| Paychex | 29.3 | 37.5 | 0.78 | 52.6 |
| State Street | 12.1 | 15.5 | 0.78 | 47.6 |
| Costco | 14.1 | 18.0 | 0.78 | 67.9 |
| General Electric | 15.0 | 15.0 | 1.00 | 44.0 |
| GATX | 11.5 | 10.5 | 1.10 | 19.7 |
| Albertsons | 17.1 | 15.5 | 1.10 | 26.6 |
| Merck | 18.4 | 14.0 | 1.31 | 33.4 |
| Chevron | 16.0 | 12.0 | 1.33 | 20.3 |
| Disney | 20.4 | 13.0 | 1.56 | 15.9 |
| Whirlpool | 19.5 | 11.0 | 1.77 | 3.8 |

Source:   Value Line and Company Reports

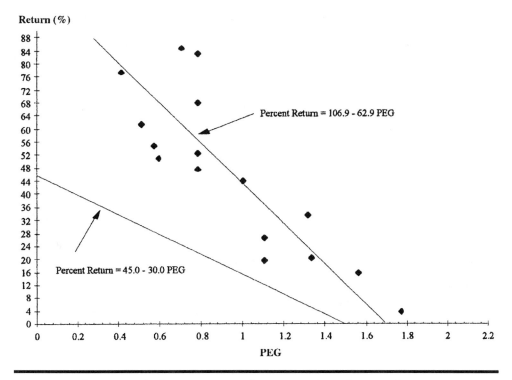

**Figure 6.2   The PEG Ratio for 1995 versus the Average Annual Return for 1995–1998**

We can try this formula to attempt to predict the future return for selected stocks for the period 1999–2002 as shown in Table 6.8.

Clearly, the best values in early 1999 appear to be out of favor stocks such as Global Marine, MGIC, and Washington Mutual. Buckle is a mid-cap stock that was not highly valued in early 1999. A great opportunity for Carlisle and Maytag is predicted by the return formula.

Historically, PEG ratios above 1.50 are warning signs. Of course, stocks with high, consistent growth rates over a five- or ten-year period should be valued highly, thus commanding a PEG greater than 1.0. Microsoft showed a PEG = 1.92 at the time of this calculation and may appear to be overvalued. Nevertheless, Microsoft has met all its growth targets for the past decade, and investors accept higher PE ratios for stocks with consistent earnings growth. Conversely, Minnesota Mining has experienced a decline in earnings from 1997 to 1998 and has flat earnings over 1996–1998. Thus, the PEG = 2.38 may be a warning, and we may expect a loss or zero return over 1999–2003. Examining these two stocks we would purchase Microsoft and avoid Minnesota Mining. Peter Lynch of Fidelity Investments used the PEG tool along with other tools in his successful tenure as a portfolio manager.

## Return on Assets

Return on assets (RoA) is a good indicator of internally generated profitability. We note that

$$RoA = E/A = E/S \times S/A.$$

**Table 6.8** **Estimation of the Annual Return for the Period 1999–2002 Using the Formula R = 45 − 30 PEG.**

| Name | March, 1999 PE | 1999–2002 Expected Earnings Growth (%) | 1999 PEG | 1999–2002 Predicted Average Annual Return (%) |
|---|---|---|---|---|
| AES | 25.2 | 25.0 | 1.00 | 15.0 |
| Airborne Freight | 14.1 | 28.0 | 0.50 | 30.0 |
| America Online | 150.0 | 100.0 | 1.50 | 0.0 |
| Biogen | 40.4 | 41.0 | 0.99 | 15.3 |
| Buckle | 14.9 | 23.5 | 0.63 | 26.1 |
| Carlisle | 14.3 | 21.0 | 0.68 | 24.6 |
| Global Marine | 8.6 | 20.5 | 0.42 | 32.4 |
| Claires Stores | 14.9 | 16.5 | 0.90 | 18.0 |
| First Union | 13.1 | 14.0 | 0.94 | 16.8 |
| Johnson Controls | 17.1 | 16.5 | 1.04 | 13.8 |
| Maytag | 17.1 | 22.5 | 0.76 | 22.2 |
| Mellon Bank | 19.6 | 17.0 | 1.15 | 10.5 |
| MGIC Investment | 9.1 | 20.0 | 0.46 | 31.2 |
| Microsoft | 58.5 | 30.5 | 1.92 | 0.0 |
| Minnesota Mining | 20.2 | 8.5 | 2.38 | 0.0 |
| Network Assoc. | 25.9 | 27.5 | 0.94 | 16.8 |
| New York Times | 21.3 | 14.5 | 1.47 | 0.9 |
| Quantum | 17.1 | 22.0 | 0.78 | 21.6 |
| Ross Stores | 15.1 | 20.0 | 0.76 | 22.2 |
| SCI Systems | 15.3 | 20.5 | 0.75 | 22.5 |
| Southwest Air | 23.5 | 17.5 | 1.34 | 4.8 |
| Sterling Software | 16.0 | 20.0 | 0.80 | 21.0 |
| United Healthcare | 17.6 | 13.5 | 1.30 | 6.0 |
| Washington Mutual | 12.8 | 24.5 | 0.52 | 29.4 |

Thus, RoA is the product of profit margin and asset turnover. Asset turnover, S/A, is a measure of how efficiently a company uses its asset base to generate revenue. As an example, consider the successful retailer, Gap, which had a 1998 profit margin of

$$PM = \frac{E}{S} \times 100\% = \frac{800}{9054} \times 100\% = 8.8\%.$$

The asset turnover for Gap is

$$Sales/Assets = 9054/3583 = 2.52.$$

Thus, Gap's return on assets is

$$RoA = PM \times S/A = 22.3\%.$$

The return on assets for selected companies along with the profit margin, asset turnover, and average annual return is given in Table 6.9. Companies with higher return on assets can generate superior returns as shown in the Table. The return on assets for selected banks, an insurance company and a financial services company are also given in Table 6.9. Financial institutions hold their depositors' funds or premiums

**Table 6.9 Return on Assets for Selected Companies (1998)**

| Name | Profits | Sales | Profit Margin (%) | Assets | Sales Assets | Return on Assets (%) | 1995–98 Average Annual Return (%) |
|------|---------|-------|-------------------|--------|--------------|----------------------|-----------------------------------|
| Microsoft | 4,786 | 14,484 | 33.0 | 14,387 | 1.01 | 33.2 | 84.9 |
| Medtronic | 710 | 3,035 | 23.4 | 2,775 | 1.09 | 25.6 | 39.2 |
| Gap | 800 | 9,054 | 8.8 | 3,583 | 2.52 | 22.3 | 83.1 |
| Cisco | 1,873 | 8,459 | 22.1 | 8,917 | 0.95 | 21.0 | 77.5 |
| Intel | 6,068 | 26,273 | 23.1 | 31,471 | 0.83 | 19.2 | 61.4 |
| Network Assoc. | 225 | 970 | 23.2 | 1,183 | 0.82 | 19.0 | 72.1 |
| Merck | 5,230 | 26,530 | 19.71 | 31,157 | 0.85 | 16.8 | 33.4 |
| Sun Microsystems | 906 | 9,791 | 9.26 | 5,711 | 1.71 | 15.9 | 55.4 |
| Sara Lee | 1,102 | 20,011 | 5.51 | 10,989 | 1.82 | 10.0 | 23.2 |
| Clorox | 298 | 2,741 | 10.87 | 3,030 | 0.91 | 9.8 | 51.3 |
| Hewlett-Packard | 3,065 | 47,061 | 6.51 | 33,673 | 1.40 | 9.1 | 18.9 |
| Applied Materials | 436 | 4,042 | 10.79 | 4,930 | 0.82 | 8.9 | 29.4 |
| Sysco | 325 | 15,328 | 2.12 | 3,780 | 4.06 | 8.6 | 21.1 |
| Allied Signal | 1,331 | 15,128 | 8.8 | 15,560 | 0.97 | 8.5 | 24.9 |
| Analog Devices | 143 | 1,231 | 11.62 | 1,862 | 0.66 | 7.7 | 21.1 |
| Alltel | 580 | 5,175 | 11.21 | 9,374 | 0.55 | 6.2 | 30.5 |
| ConAgra | 628 | 23,841 | 2.63 | 11,703 | 2.04 | 5.4 | 17.6 |
| Qualcomm | 126 | 3,348 | 3.75 | 2,567 | 1.30 | 4.9 | 7.2 |
| Minnesota Power | 95 | 1,050 | 9.05 | 2,317 | 0.45 | 4.1 | 23.2 |

**Banks, Insurance Companies, and Financial Companies**

| Name | Profits | Sales | Profit Margin (%) | Assets | Sales Assets | Return on Assets (%) | 1995–98 Average Annual Return (%) |
|------|---------|-------|-------------------|--------|--------------|----------------------|-----------------------------------|
| Amer. Int'l Group | 3,766 | —— | —— | 194,000 | —— | 1.94 | 33.4 |
| Fifth Third Bank | 555 | —— | —— | 28,922 | —— | 1.92 | 50.8 |
| TCF Financial | 155 | —— | —— | 9,744 | —— | 1.59 | 15.8 |
| First Tennessee | 226 | —— | —— | 16,721 | —— | 1.35 | 39.6 |

and earn an investment return on these moneys. When we compare leading firms in an industry, we will record and compare the return on assets.

## The Discounted Earnings Model

If we assume that the value of a stock rises along with a rise in earnings then we can formulate a discount model that values a share of stock based on the estimated flow of earnings (or, alternatively, cash flow—see Chapter 7). Then the value, or price, today should be

$$\text{Price} = \frac{E_1}{(1+r)^1} + \frac{E_2}{(1+r)^2} + \cdots + \frac{E_n}{(1+r)^n} + \cdots$$

where r is the discount rate and $E_n$ is the earnings over the period n. We assume a growth rate of earnings, g, so that

$$\text{Price} = \sum_{n=1}^{N} E_1 \frac{(1+g)^n}{(1+r)^n}.$$

If we assume that we will hold the stock in perpetuity (infinite time—only mathematically feasible) we have the price earnings ratio, Price/E$_1$,

$$\frac{\text{Price}}{\text{Earnings}} = \sum_{n=1}^{\infty} x^n \tag{6.2}$$

where

$$x^n = (1 + g)^n/(1 + r)^n.$$

It can be shown that the equation becomes a special form when g is less than r as follows:

$$PE = \frac{1}{r - g}. \tag{6.3}$$

This model assumes a stable, steady growth rate in perpetuity (forever). This formula can only be used with low growth rate companies like utilities and then only with caution. If we assume a utility can continually grow its earnings at 3%, then, g = 0.03. The discount rate for a slow growth, lower risk company might be 10% and r = 0.10. Then we have

$$PE = \frac{1}{0.10 - 0.03} = \frac{1}{0.07} = 14.$$

Let's consider the S&P 500 in slow growth times and assume g = 0.05 and the discount rate is r = 0.11. Then the S&P 500 PE in modest times might be

$$PE = \frac{1}{0.11 - 0.05} = \frac{1}{0.06} = 16.7.$$

In times of high inflation and high interest rates the S&P 500 growing at 5% may warrant a discount rate of 13 percent. Then the S&P 500 PE would be

$$PE = \frac{1}{0.13 - 0.05} = \frac{1}{0.08} = 12.5.$$

Inflation, interest rates, and expected risk cause the analyst to adjust the discount rate, r. When inflation and interest rates exceed 5% and 10% respectively, the discount rate may be as high as r = 0.15.

Let's develop a more realistic model using a limited period of N years with an expected growth rate, g. We will assume that the discount rate is r. Then the PE ratio is

$$PE = \sum_{n=1}^{N} x^n + \frac{TV}{(1 + r)^N}$$

where the **terminal value**, TV, is also discounted. If we assume a steady (perpetual) growth rate, g$_p$, for the period after the N years we can use

$$TV = \frac{(1 + g_p)(1 + g)^N}{r - g_p}$$

where the perpetual growth rate is $g_p$. Then we have

$$PE = \sum_{n=1}^{N} x^n + \frac{x^N(1 + g_p)}{(r - g_p)}.$$

As an example, consider the case where $g = r$ so that $x = 1$. Also let $g_p = 0$. Then

$$PE = N + \frac{1}{r}.$$

If $N = 5$ and $r = 0.10$, we obtain $PE = 15$.

This sample model assumes a visibility of earnings over five years and then zero earnings growth after the five-year period. There are very few companies for which we are able to estimate a growth rate beyond five years. The Value Line Survey provides an estimate of the growth rate for a three- to five-year horizon. Now let's use $g_p = 0$ and $N = 5$ with more general rates of initial growth. Therefore, we have

$$PE = \sum_{n=1}^{5} x^n + \frac{x^5}{r}. \qquad (6.4)$$

For $r = 0.10$ and $g = 0.22$ we have $x = 1.109$. Then

$$PE = 1.109 + 1.230 + 1.364 + 1.513 + 1.678\left(1 + \frac{1}{0.10}\right) = 23.68.$$

Using Equation 6.4 we proceed to develop the value of PE for a range of g and r as shown in Table 6.10. A reasonable approximation to the PE in Table 6.10 is the use of the formula

$$PE = (G + 2)(6/I) \qquad (6.5)$$

where G is the percent growth rate of earnings and I is the yield of the 30-year Treasury Bond. The discount rate, r, may be approximated by a 5% premium added to the 30-year Treasury Bond yield. Thus, when the T-Bond yield is 5.0%, the discount rate, r, is 10.0%. Thus the appropriate PE using Equation (6.5) for State Street Corp. is

$$PE = (21.5 + 2)(6/5) = 28.2.$$

The actual PE for State Street on March 8, 1999 (Value Line) was 27.2. If you determine the expected PE from Table 6.10 for $g = 0.22$ and $r = 0.10$, you obtain PE = 23.7. Modest changes on r or g will modify the expected PE. If g drops to 0.18 and r increases to 0.11 due to a shift upward to 6.0% for the yield of the Treasury Bond, then, using Table 6.10, we expect PE to decline to 18.37. Using the formula of equation (6.5) we get

$$PE = (18 + 3)(6/6) = 20.$$

## Conclusion

Useful measures of valuation for a company can be used together to determine if a firm is undervalued and should be considered for purchase. We seek to select great

**Table 6. 10   The Calculated PE Ratio Using the Discounted Earnings Model of Equation (6.4)**

| g* \ r** | 0.09 | 0.10 | 0.11 |
|---|---|---|---|
| 0.00 | 11.11 | 10.00 | 9.09 |
| 0.02 | 12.09 | 10.87 | 9.86 |
| 0.04 | 13.14 | 11.79 | 10.69 |
| 0.06 | 14.27 | 12.79 | 11.58 |
| 0.08 | 15.48 | 13.86 | 12.54 |
| 0.09 | 16.11 | 14.42 | 13.04 |
| 0.10 | 16.77 | 15.00 | 13.56 |
| 0.11 | 17.45 | 15.60 | 14.09 |
| 0.12 | 18.16 | 16.22 | 14.65 |
| 0.14 | 19.64 | 17.53 | 15.81 |
| 0.16 | 21.22 | 18.92 | 17.05 |
| 0.18 | 22.91 | 20.41 | 18.37 |
| 0.20 | 24.70 | 21.99 | 19.78 |
| 0.22 | 26.62 | 23.68 | 21.28 |
| 0.24 | 28.66 | 25.47 | 22.87 |
| 0.26 | 30.82 | 27.37 | 24.57 |
| 0.28 | 33.16 | 29.40 | 26.36 |
| 0.30 | 35.56 | 31.54 | 28.27 |

*g = growth rate of earnings for the next five years
**r = discount rate

companies using the criteria indicated in Exhibit 6.4. When considering companies in a selected industry we will use comparisons of two or three companies and contrast the companies on each criteria.

**Exhibit 6.4   Measures of Valuation for Comparing Two or More Companies in a Selected Industry**

- A clear and understandable business franchise
- Commitment to performance and a core ideology
- Consistent earnings growth (greater than 11%)
- Low or moderate leverage (assets/equity and debt/capital)
- High sales/assets turnover
- High profit margins
- High operating margins
- High return on equity
- High return on assets
- Reasonable PE ratio for the growth rate and bond yield

# 7 Power and Value

## This chapter covers

- Cash Flow
- Price to Cash Flow Ratio
- Economic Value Added
- The Power Equation
- Price Growth Persistence, Price Stability Indices, and Safety Rating
- The Value Index
- Intrinsic Value

The **cash flow** of a firm is the net cash earned during a specific period, usually one year. To determine cash flow, you add back the noncash charges to the earnings. Using the Value Line Survey we can determine the cash flow by adding depreciation to profits. Cash flow per share is also shown on each Value Line page. For example, Harley-Davidson (HDI) had a net profit of $213.5 million and a depreciation charge of $87.4 million for a 1998 total cash flow of $301 million. Using the number of shares, 153 million, we obtain $1.97 cash flow per share as listed on the Value Line page for HDI as shown in Figure 7.1. Depreciation is identified as item 1, net profit is item 2, and cash flow per share is item 3 shown in Figure 7.1.

If a company has a positive cash flow, it will have cash to make acquisitions or invest in new activities. If it is negative, the company will have to raise capital to keep growing. We seek companies with positive, growing cash flow. The cash flow for Harley-Davidson was positive for the past decade and it grew at an average rate of 19.5% for the 10-year period through 1998. With positive cash flow the firm is able to pay all near-term obligations such as interest, rent, and the like.

The cash flow data is a reliable indicator of a company's performance. Earnings are less reliable because of the use of write-offs, one-time special charges and extraordinary items as charges against earnings. Since earnings are subject to accounting rules and extraordinary charges, it is often better to use the more reliable cash flow figures. Cash flow is undistorted by differences across industries. Earnings for Time Warner during the period 1990–1998 were consistently negative while the cash flow was always positive. If you only looked at earnings, you might have avoided this stock. In 1998, Time Warner had a profit of $168 million and a depreciation charge of $1,178 million.

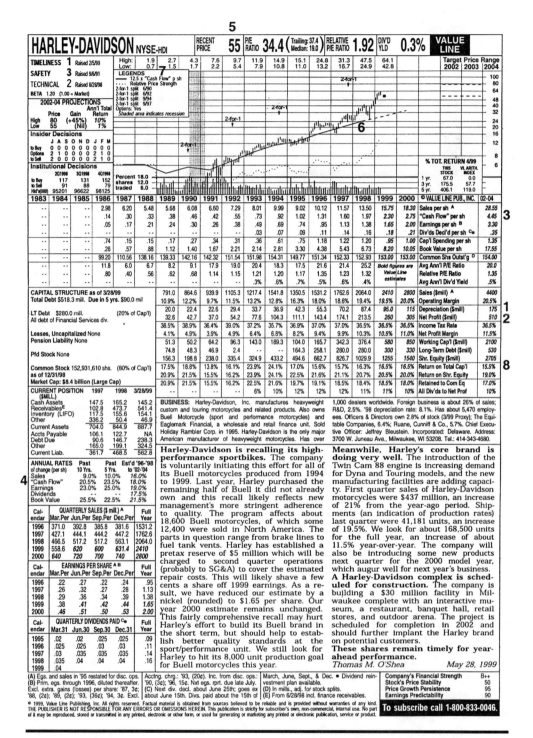

**Figure 7.1    The Value Line Survey Page for Harley-Davidson**

However, after charges, earnings were $–0.31 per share while cash flow was $0.70 per share. Time Warner provided an average annual return of 23.9% for the five-year period 1993–98.

## Price to Cash Flow Ratio

The price to earnings ratio, PE, is less reliable due to the various charges to earnings. The better reliability of cash flow leads us to consider using the forward price to cash flow ratio as:

$$PCF = \frac{Price}{Cash\ Flow}.$$

We can use a discounted cash flow model (see Equation (6.4) for the model for earnings) to obtain the equation

$$PE = \sum_{n=1}^{N} x^n + \frac{x^N}{r}. \tag{7.1}$$

where the time horizon for cash flow growth is N years, $x = (1 + g) / (1 + r)$, g = projected growth rate of cash flow, and r = discount rate. We assume that we can project growth for five years and thus N = 5. Given a growth rate g and a discount rate r, one can use Table 7.1 to estimate the PCF ratio. Table 7.1 is a more complete version of Table 6.10 developed in Chapter 6. We use a discount rate r = 1 + 0.05, where I = interest rate of 10-year Treasury Bond. Consider Harley-Davidson with a projected growth rate of cash flow equal to 18.0% as shown as item 4 in Figure 7.1. Using an interest rate I = 0.05 we have r = 0.10 and g = 0.180. Using Table 7.1, we estimate a PCF ratio equal to 20.4. This is a forward ratio, providing a ratio using the current year cash flow. Value Line gives an estimated 1999 cash flow of $2.30 as shown in line item 3. Thus, the intrinsic value at the time of this calculation, June 1999, was

$$Intrinsic\ Value = PCF \times CF = 20.4 \times \$2.30 = \$46.92.$$

The actual price in June 1999 was $55.00 as shown in item 5 in Figure 7.1. Examining page 1,779 of the Value Line Survey of May 28, 1999 we note that a "Value Line" is provided as a PCF line (Item 6 in Figure 7.1) that correlates with the PCF of the past decade for Harley-Davidson. This "Value Line" shows a correlation with a PCF = 12.5 shown as item 7. Thus, the market in 1999 is valuing Harley-Davidson with a higher ratio then it did in the years 1993–1996. The increase in PCF over the years 1992–1999 is in part due to the decrease in the interest rate. During that period the 30-year T-bond yield dropped from a yield of 7% to a yield of 5%. Thus, the discount rate was r = 0.05 + 0.07 = 0.12 in 1992. Then examining Table 7.1 for g = 0.18 we obtain PCF = 16.7. These results are summarized in Table 7.2. Note how the market awarded HDI an increasing PCF as the interest rate declined.

We repeat the series of calculations of PCF for Donaldson, Allied Signal, Microsoft, and WalMart. For example, Microsoft has a Value Line correlation line with PCF = 30. In the 1992–1997 period the growth rate of cash flow was about 36%. Using the 1992 discount rate of 0.12 we determine from Table 7.1 that PCF = 31.3. For 1999 we use a projected growth rate of 27.0% and a discount rate of 10%. Then from Table 7.1 we find that PCF = 28.4. The actual PCF on March 5, 1999, was

$$PCF = \frac{\$76}{1.50} = 50.7$$

when the estimated 1999 cash flow/share was $1.50. (See Table 7.2)

**Table 7.1 The Calculated PCF Ratio Using the Discounted Cash Flow Model of Equation (7.1) with N = Five Years**

| g* \ r** | 0.09 | 0.10 | 0.11 | 0.12 |
|---|---|---|---|---|
| 0.00 | 11.11 | 10.00 | 9.09 | 8.33 |
| 0.02 | 12.09 | 10.87 | 9.86 | 9.03 |
| 0.04 | 13.14 | 11.79 | 10.69 | 9.78 |
| 0.06 | 14.27 | 12.79 | 11.58 | 10.58 |
| 0.08 | 15.48 | 13.86 | 12.54 | 11.44 |
| 0.09 | 16.11 | 14.42 | 13.04 | 11.89 |
| 0.10 | 16.77 | 15.00 | 13.56 | 12.35 |
| 0.11 | 17.45 | 15.60 | 14.09 | 12.84 |
| 0.12 | 18.16 | 16.22 | 14.65 | 13.33 |
| 0.14 | 19.64 | 17.53 | 15.81 | 14.38 |
| 0.16 | 21.22 | 18.92 | 17.05 | 15.49 |
| 0.18 | 22.91 | 20.41 | 18.37 | 16.68 |
| 0.20 | 24.70 | 21.99 | 19.78 | 17.95 |
| 0.22 | 26.62 | 23.68 | 21.28 | 19.29 |
| 0.24 | 28.66 | 25.47 | 22.87 | 20.72 |
| 0.26 | 30.82 | 27.37 | 24.57 | 22.24 |
| 0.28 | 33.16 | 29.40 | 26.36 | 23.85 |
| 0.30 | 35.56 | 31.54 | 28.27 | 25.55 |
| 0.32 | 38.15 | 33.81 | 30.28 | 27.36 |
| 0.34 | 40.89 | 36.22 | 34.42 | 29.27 |
| 0.36 | 43.79 | 38.77 | 34.68 | 31.29 |
| 0.38 | 46.86 | 41.46 | 37.07 | 33.43 |
| 0.40 | 50.11 | 44.31 | 39.60 | 35.69 |

* g = Growth rate of earnings for the next three to five years.
**r = Discount rate.

**Table 7.2 The Calculated PCF Ratio Using the Discounted Cash Flow Model of Equation (7.1) with N = Five Years**

| Name | Value Line Correlation Line | Table 7.1 Using Projected Growth Rate for: | | Actual Calculated March 5, 1999 |
|---|---|---|---|---|
| | | 1992–97 | 1999–2003 | |
| Allied Signal | 11.0 | 12.4 | 15.6 | 10.7 |
| Central News | 11.0 | 17.6 | 17.5 | 10.5 |
| Donaldson | 10.0 | 15.0 | 12.8 | 10.0 |
| Harley-Davidson | 12.5 | 16.7 | 20.8 | 20.9 |
| Microsoft | 30.0 | 31.3 | 28.4 | 50.7 |
| WalMart | 16.0 | 16.0 | 18.9 | 24.2 |

The most conservative valuation is provided by the "Value Line" PCF. Allied Signal and Central News had actual values on March 5, 1999 that were essentially the same as the Value Line figure. Microsoft had an actual PCF = 50.7 while its Value Line PCF = 30.0 The PCF of any company is a function of the growth rate of cash flow, g, the discount rate, r, the market's perception of the companies ability to sustain or improve its growth rate, and the demand for shares of stock by investors. The last two factors are not factored into the Table 7.1. Microsoft is a company that has sustained and improved its growth rate, and demand for its shares is high.

The effect of a great record of sustainability and demand for a company stock is imbedded in the length of time, N, that the purchaser believes is appropriate for use in Equation (7.1). For Microsoft, let us assume that an eight-year projection of cash flow growth is possible. Using Equation (7.1) with N = 8 we have

$$PCF = \sum_{n=1}^{8} x^n + \frac{x^8}{r}.$$

and g = 0.27 and r = 0.10 for 1999. Then x = 1.155 and we calculate PCF = 47.7. The longer period one can project the earnings growth is sustainable the higher is the resulting PCF ratio. This calculation indicates that investors are projecting Microsoft's growth over an eight-year period.

We can also define a ratio of PCF to cash flow growth rate. We call this ratio PCFG, where PCF is the price to cash flow and G is the percent growth rate of cash flow. The PCFG ratio is analogous to the PEG ratio for earnings. Microsoft has a projected growth rate in 1999 of 27% and an estimated PCF for 1999 of 50.7. Therefore,

$$PCFG = \frac{PCF}{G} = \frac{50.7}{27} = 1.88.$$

Undervalued stocks may have a PCFG less than or equal to 1.0. Overvalued stocks may have a PCFG greater than 2.0.

## Economic Value Added

Over the past decade there has been an increasing tendency to use alternative accounting methods, adopt one-time charges and other factors which decrease the reliability of earnings. In addition, as the number of write-downs occur the equity capital of a firm is reduced, thus increasing the return on equity. Furthermore, as we shift from traditional industrial companies to those dependent on intellectual capital we need to insure that managers are working to increase shareholder value. In the cash-flow-based value analysis, $1 million spent for knowledge and $1 million spent for tangible assets are treated identically. Thus, we want a cash flow analysis incorporated within the calculation of intrinsic value.

The appropriate discount rate is the weighted costs of debt and equity capital. If a company has 30% of its capital from debt (after tax) and 70% from equity, then it may calculate its **cost of capital** (CC) as

$$CC = 0.3 \, (6.0\%) + 0.7 \, (11\%) = 9.5\%$$

where the after-tax cost of debt is 6% and the cost of equity capital is 11% for this company. For a company with greater risk the cost of equity capital may be 12% or 13% [Rappaport, 1998].

The management of a company should be focused on a measure called **economic value added** (EVA) which is operating profits less the cost of the capital employed to produce those earnings [Stewart, 1991]. EVA can be made to grow if capital can be invested in projects that earn more than the cost of capital. Thus, a company that earns a **return on capital** (Rcap) greater than the cost of capital (CC) creates value for its shareholders. We define the return on capital (Rcap) as

$$Rcap = \frac{Profit}{Total\ Capital} = \frac{PFT}{TC}$$

where total capital = debt + equity. The Value Line Survey provides the return on capital for each of its 1,700 companies. The actual rate of return on capital is a key performance measure for any company.

The economic value added (EVA) of a company is

$$EVA = (Rcap - CC) \times Total\ Capital \qquad (7.2)$$

Thus, value is created by the spread, S, where

$$S = Rcap - CC$$

and the **market value added** (MVA) is

$$MVA = Market\ Value - Capital$$
$$= Present\ Value\ of\ All\ Future\ EVA.$$

Logically, we seek companies that create MVA by creating positive EVA. Thus, we seek companies with a high return on capital and a lower cost of capital. Typically, a firm finds its optimal cost of capital is achieved with a debt-to-capital ratio greater than 10% and less than 50%. As the debt to capital ratio is increased, the risk from fixed interest costs increase and the investor demands a higher interest rate on the debt and a higher cost of equity capital.

Returning to Equation (7.2) we note that the Market Value (price × number of shares) of a firm will increase each year as long as the EVA remains positive year after year. Table 7.3 shows the MVA and the return on capital for three large retail stores. Note that Wal-Mart created a market value added of $150 billion by maintaining a return on capital of 14.1% while K-Mart destroyed $2.5 billion of market value added by having a return on capital less than its cost of capital.

Return on capital is

$$Rcap = \frac{PFT}{TC} = \frac{PFT}{S} \times \frac{S}{TC} = Profit\ Margin \times \frac{Sales}{Capital}.$$

**Table 7.3  The MVA and the Return on Capital for Three Large Retail Stores**

| Name | Market Value ($ in bil.) 2/99 | Capital ($ in bil.) 1998 | MVA ($ in bil.) | Rcap 3-Year Average (%) | Estimated Cost of Capital (%) | Spread = Rcap − CC (%) |
|---|---|---|---|---|---|---|
| Wal-Mart | $180.0 | $29.8 | $150.2 | 14.1 | 9.9 | 4.2 |
| Costco | 16.8 | 3.9 | 12.9 | 10.9 | 9.9 | 1.0 |
| K-Mart | 7.9 | 10.4 | −2.5 | 5.6 | 10.1 | −4.5 |

In order to increase the return on capital we need to improve the profit margin and the sales generated for a fixed capital investment.

The EVA for the top 100 companies are published each year by *Fortune* magazine in their November issue providing the results for the proceeding year. Table 7.4 shows the EVA for selected companies for 1997 as well as the return on capital, cost of capital, debt/capital and the three-year average return for 1995–98. Note that the companies with the higher return on capital also provided the higher average annual return. Compare the two auto companies on the basis of the spread between the return on capital and the cost of capital. Ford created wealth in 1997 and General Motors destroyed shareholder wealth in 1997 according to these figures reported in *Fortune* magazine [Tully, 1998]. The 1998 return on capital reported by Value Line for Ford and General Motors was 8.0% and 7.5% respectively. Assuming the cost of capital for both was approximately 7.5%, then Ford was again the creator of wealth.

The cost of capital for a utility or a REIT is in the range of 7% to 9%. Utility companies typically have a large debt/capital ratio. The return on capital, estimated cost of capital, and debt/capital ratio is given in Table 7.5 for four companies.

**Table 7.4  EVA, Return on Capital for 1997 and Average Annual Return for 1995–98 for Selected Stocks**

| Name | EVA ($ in Mil.) | Return on Capital (%) | Cost of Capital (%) | Spread (%) | Debt/ Capital (%) | Average Annual Return 1995–98 (%) |
|---|---|---|---|---|---|---|
| Microsoft | $2,781 | 52.9 | 14.2 | 38.7 | 4 | 84.9 |
| Cisco | 1,472 | 50.9 | 14.3 | 36.6 | 0 | 77.5 |
| Intel | 4,821 | 42.7 | 15.1 | 27.6 | 3 | 61.4 |
| Schering Plough | 4,970 | 36.0 | 14.0 | 22.0 | 1 | 61.8 |
| Bristol Meyers | 1,802 | 25.3 | 12.5 | 12.8 | 15 | 49.6 |
| Abbott Labs | 1,200 | 23.2 | 10.9 | 12.3 | 20 | 35.4 |
| Gillette | 618 | 18.8 | 12.0 | 6.8 | 29 | 23.7 |
| Home Depot | 462 | 16.8 | 10.8 | 6.0 | 13 | 57.2 |
| Procter & Gamble | 587 | 15.2 | 12.8 | 2.4 | 34 | 32.1 |
| Ford | 3,089 | 14.4 | 9.1 | 2.3 | 73 | 51.8 |
| General Motors | −4,120 | 4.4 | 9.4 | −5.0 | 69 | 16.3 |

**Source:** *Fortune,* Nov. 9, 1998, pg. 193.

**Table 7.5  The Return on Capital, Cost of Capital, Spread, and Debt to Capital for Four Companies**

| Name | Industry | 1998 Return on Capital (%) | Estimated Cost of Capital (%) | Spread (%) | 1998 Debt/Capital (%) |
|---|---|---|---|---|---|
| Duke Energy | Utility | 9.5 | 8.5 | 1.0 | 49 |
| Teco Energy | Utility | 10.5 | 9.0 | 1.0 | 42 |
| New Plan Excel | REIT | 9.0 | 8.5 | 0.5 | 37 |
| Philadelphia Suburban | Water Utility | 7.5 | 7.0 | 0.5 | 55 |

**Source:** Value Line Survey.

Clearly, the return on capital is the best measure of wealth creation and shareholder value. The Value Line Survey provides the estimated return on capital for the current and future years as shown by item 8 in Figure 7.1. However, Value Line does not provide data on the cost of capital. The cost of capital is at least 9% for industrial companies and may be less than 9% for utilities and REITs because their debt/capital ratio is 50% or greater. Since we do not know the cost of capital for each company we will calculate the intrinsic value of a company based, in part, on the estimated return on capital.

## The Power Equation

As a result of our earlier discussion, we assert that the intrinsic value of a company's stock is related to its return on capital and its growth in cash flow, assuming the return on capital exceeds the cost of capital. We also assert that the intrinsic value is directly related to the perceived sustainability of this return on capital and growth rate of cash flow. Thus we assert that the intrinsic value can be represented as

$$\text{Intrinsic Value} = \text{Function (Rcap, G, Sus)}$$

where Rcap = return on capital in percent, G = growth rate of cash flow in percent, and Sus = sustainability of the performance.

We are interested in selecting stocks that have the power to adapt to changes in the competitive marketplace and sustain excellent returns for their shareholders. The market is dynamic and we can resort to a physical analogy of a dynamic process. The power exerted on a physical object is the product of a sustainable force, $f$, and its velocity or

$$p = f \times v.$$

In a similar manner we can describe the **power** of a company's stock in terms of the three factors Rcap, G, and Sus as

$$\text{Power} = \text{Rcap} \times \text{G} \times \text{Sus} \tag{7.3}$$

We can test this power model for selected stocks over the five-year period 1993–98 and attempt to demonstrate the utility of the power equation (Equation (7.3)).

First, we need to devise a measure of the sustainability of a company's power through varying market conditions. A company's ability to sustain its power can be determined, in part, using several measures reported by Value Line. In the box in the lower right hand corner of each Value Line page, four sustainability (or strength) measures are provided: Financial Strength, Price Stability, Price Growth Persistence, and Earnings Predictability. The definitions of these four measures are given in Exhibit 7.1. For industrial and financial companies we use Financial Strength and Price Growth Persistence to calculate Sus. For slow growing utilities and REITs we use Financial Strength and the Price Stability Index to calculate Sus.

Financial Strength Rating is given in letters: A, B, and C, and we devise a table to convert these letters to numbers as shown in Table 7.6. Then, we can define **sustainability** as

**Exhibit 7.1 The Definition of Four Value Line Sustainability Measures**

**Financial Strength Rating**—A relative measure of financial strength of the companies reviewed by Value Line. The relative rating range from A$^{++}$ (strongest) down to C (weakest), in nine steps.

**Price Stability Index**—A measure of the stability of a stock's price. It includes sensitivity to the marker (Beta) as well as the stock's inherent volatility. Value Line Stability ratings range from 100 (highest) to 5 (lowest).

**Price Growth Persistence Index**—Measures the historic tendency of a stock to show persistent price growth compared to the average stock for an eight-year period. Value Line Persistence ratings range from 100 (highest) to 5 (lowest).

**Earnings Predictability Index**—A measure of the reliability of an earnings forecast. Predictability is based upon the stability of year-to-year comparisons, with recent years being weighted more heavily than earlier ones. The most reliable forecasts tend to be those with the highest rating (100); the least reliable, the lowest (5). The earnings stability is derived from the standard deviation of percentage changes in quarterly earnings over an eight-year period.

**Table 7.6   Financial Strength Conversion**

| Financial Strength Rating | B | B$^+$ | B$^{++}$ | A | A$^+$ | A$^{++}$ |
|---|---|---|---|---|---|---|
| Equivalent Number | 0.90 | 1.0 | 1.1 | 1.2 | 1.3 | 1.4 |

$$\text{Sus} = \text{Financial Strength Rating} \times \frac{\text{Price Growth Persistence}}{100}$$

where we use Financial Strength Rating equivalent number and the Price Growth Persistence. Then, for industrial and financial companies we have

$$\text{Power} = \text{Rcap} \times \text{G} \times \text{Fin. Str.} \times \frac{\text{Price Growth Persistence}}{100}$$

Let us calculate the power for Harley-Davidson using the 1999 numbers in the Value Line page shown in Figure 7.1. Then, we have

$$\text{Power} = \text{Rcap} \times \text{G} \times \text{Fin. Str.} \times \frac{\text{Pr. Gr. Persis.}}{100}$$

$$= 16.5 \times 18.0 \times 1.1 \times \frac{95}{100} = 310.4.$$

The calculation of the Power for Microsoft for the past five-year period is shown in Table 7.7 where we use the average RCap over that period. We also show the average annual return for that period (1993–98). Value Line provides an estimate of the RCap for 1999 and an estimate for 2001–03 as 31% and 25% respectively. We use the average of these two numbers for our forward projections (Rcap = 28%). Value Line gives a projected growth rate G of 27.0%. Using these numbers and assuming Microsoft's Financial Strength and Price Growth Persistence remain steady we calculate the Power for the next five years as shown in Table 7.7. Since the projected Power is 0.73 of the past power, we can estimate an average annual return of 0.73 times the past return to estimate a return of 50%. The accuracy of this projection is limited but we can readjust our projections as Value Line issues updates on projected Rcap, G, or Fin. Str.

**Table 7.7   Power Calculation**

| | Rcap × | G × | Fin. Str. × | Pr. Gr. Persis. / 100 | = Power | Average Annual Return(%) |
|---|---|---|---|---|---|---|
| | (%) | (%) | | | | |
| **Past** | | | | | | |
| Five Years 1993–98 | 29.3 | 35.5 | 1.40 | 0.95 | 1383.4 | 68.9 |
| **Projected** | | | | | | |
| Next Five Years 1999–2004 | 28.0 | 27.0 | 1.40 | 0.95 | 1005.5 | 50.1 |

The Power of 22 stocks over the period 1993–98 and the average annual return of these stocks are given in Table 7.8. Note that we use Price Stability instead of Price Growth Persistence for utilities—in this case American Electric Power. Also, we adjusted the Price Growth Persistence for Sun Microsystems to the past five-year period. Clearly, higher returns were obtained by companies with higher Power over the five-year period. We assert the predictability of future returns based on the power of the company.

**Table 7.8   The Calculation of Power and the Average Annual Return for 22 Stocks for the Period 1993–98**

| Name | Average Rcap × (%) | G × (%) | Fin. Str. × | Pr. Gr. Persis. / 100 | = Power | Average Annual Return 1993–98 (%) |
|---|---|---|---|---|---|---|
| Dell | 48.5 | 48.0 | 1.40 | 0.95 | 3096.2 | 152.9 |
| Network Associates | 27.4 | 58.0 | 1.10 | 0.85 | 1485.9 | 113.9 |
| Microsoft | 29.3 | 35.5 | 1.40 | 0.95 | 1383.4 | 68.9 |
| Cisco | 32.7 | 61.0 | 1.40 | 1.00 | 1144.5 | 66.8 |
| Schwab | 20.6 | 35.0 | 1.0 | 1.00 | 721.0 | 64.7 |
| Sun Microsystems | 23.5 | 20.6 | 1.30 | 0.95 | 597.9 | 63.7 |
| Tellabs | 26.8 | 56.5 | 1.10 | 1.00 | 1665.6 | 63.3 |
| EMC | 22.6 | 63.5 | 1.30 | 0.95 | 1772.3 | 59.4 |
| Pfizer | 15.0 | 26.1 | 1.40 | 0.85 | 465.9 | 51.2 |
| Intel | 29.0 | 41.5 | 1.40 | 0.95 | 1600.6 | 50.6 |
| Gap | 20.0 | 28.9 | 1.30 | 0.70 | 526.0 | 46.3 |
| Danaher | 13.4 | 24.0 | 1.10 | 1.00 | 353.8 | 41.9 |
| Home Depot | 14.3 | 26.5 | 1.40 | 0.80 | 424.4 | 36.5 |
| General Electric | 23.0 | 10.0 | 1.40 | 1.00 | 322.0 | 34.2 |
| Cintas | 14.1 | 15.5 | 1.10 | 0.95 | 228.4 | 33.6 |
| Interpublic | 17.5 | 12.0 | 1.30 | 0.95 | 259.4 | 32.0 |
| Auto Data Process | 16.9 | 13.5 | 1.40 | 0.95 | 303.4 | 24.9 |
| Emerson Electric | 18.1 | 10.5 | 1.40 | 1.00 | 266.1 | 17.6 |
| American Electric Power | 7.8 | 2.5 | 1.20 | 1.00 | 23.4 | 11.5 |
| Minnesota Mining | 20.3 | 3.5 | 1.40 | 0.75 | 74.6 | 9.4 |
| Penney, J. C. | 9.2 | 6.0 | 1.10 | 0.50 | 30.4 | 1.5 |
| Toys R Us | 10.6 | 8.5 | 1.20 | 0.25 | 27.0 | −16.2 |

## Price Growth Persistence, Price Stability Indices, and the Value Line Safety Rating

The Value Line Price Growth Persistence Index measures a stock's tendency to exhibit persistent growth compared to the average stock. The Price Stability Index measures the stability of the stock price. The Safety Rank is a measure of the potential risk of a stock with rank 1 being safest. Table 7.9 shows these three measures for 24 well-known stocks. We show these data points on Figure 7.2. The first quadrant with high Price Growth Persistence and low Price Stability contains the stocks with high growth of cash flow and high returns. Quadrant 1 contains Gazelle stocks. Quadrant 2 holds stocks with high Price Growth Persistence and high Price Stability and better Safety ranks (1 or 2). This quadrant holds the Hare stocks. Quadrant 3 holds REIT and utility (Tortoise) stocks with high Price Stability but relatively low Price Growth Persistence. Finally, Quadrant 4 holds stocks with low Price Growth and Price Stability. The Quadrant 4 stocks may be candidates for the Contra portfolio.

In Quadrant 2, Firstar and General Electric have very high price growth persistence and very high price stability. These two firms provide high consistent returns because investors count on them to provide very consistent growth over the long term.

**Table 7.9    The Calculation of Power and the Average Annual Return for 22 Stocks for the Period**

| Number | Name | Price Growth Persistence | Price Stability | Value Line Safety | 1993–98 Average Annual Return (%) | Quadrant |
|--------|------|--------------------------|-----------------|-------------------|-----------------------------------|----------|
| 1 | Dell | 95 | 10 | 3 | 152.9 | |
| 2 | EMC | 95 | 15 | 3 | 59.4 | **1** |
| 3 | Sun Microsystems | 75 | 20 | 3 | 63.7 | |
| 4 | Intel | 95 | 35 | 3 | 50.6 | |
| 5 | Gap | 70 | 30 | 3 | 46.3 | **Gazelles** |
| 6 | Cisco | 100 | 20 | 3 | 66.8 | |
| 7 | Microsoft | 95 | 55 | 2 | 68.9 | |
| 8 | Firstar | 95 | 95 | 2 | 63.7 | **2** |
| 9 | Amer. Int. Group | 100 | 75 | 2 | 30.5 | |
| 10 | Pfizer | 85 | 65 | 2 | 51.2 | |
| 11 | Walgreen | 90 | 70 | 2 | 43.6 | |
| 12 | SBC Commun. | 85 | 80 | 2 | 24.8 | |
| 13 | Home Depot | 80 | 60 | 2 | 36.5 | **Hares** |
| 14 | General Electric | 100 | 90 | 1 | 34.2 | |
| 15 | DuPont | 95 | 80 | 1 | 20.2 | |
| 16 | TNP | 20 | 80 | 3 | 22.7 | |
| 17 | United Water | 5 | 85 | 3 | 18.3 | **3** |
| 18 | New Plan Realty | 20 | 95 | 2 | 16.3 | |
| 19 | Amer. Electric | 25 | 100 | 2 | 11.5 | |
| 20 | CMS Energy | 30 | 100 | 3 | 17.9 | **Tortoises** |
| 21 | Nicor | 40 | 100 | 1 | 13.7 | |
| 22 | Calgon Carbon | 5 | 40 | 3 | −5.9 | **4** |
| 23 | Navistar | 5 | 25 | 4 | 0.6 | |
| 24 | K-Mart | 15 | 25 | 4 | 0.0 | **Contras** |

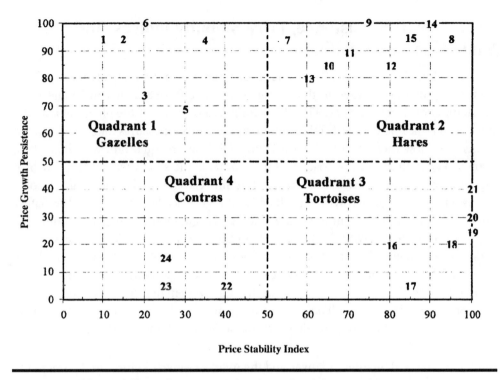

**Figure 7.2    Price Stability Index versus Price Growth Persistence**

## The Value Index

The power of a stock is an excellent measure of the ability of the stock to deliver attractive future returns. The **Value Index** of the stock is

$$\text{Value Index} = \frac{\text{Power}}{\text{Estimated PCF}}$$

where the **estimated PCF** is the current price of the stock divided by the expected cash flow over the current year. Thus, if a stock has a price today (1999) of $30.00 and the Value Line estimate of the 1999 cash flow is $2.40, we have

$$\text{Est. PCF} = 12.5.$$

If the stock has a power of 340, then the value index is

$$\text{VI} = 340/12.5 = 27.2.$$

Stocks with higher value indexes (greater than 15) will provide higher potential annual returns.

If later in the same year the price rises to $40.00, we have an estimated PCF equal to

$$\text{Est. PCF} = \$40.00/2.40 = 16.7.$$

Then, the Value Index drops to

$$\text{VI} = 340/16.7 = 20.4.$$

The Power and Value Index for four firms in each of four industries is shown in Table 7.10. General Electric has the highest power in the electrical equipment industry. However, with a PCF = 24.7 at the time of this writing, we determine the VI = 20.1. On the other hand, Johnson Controls has a more modest power of 189 but also a lower current PCF of 6.6. Thus, its Value Index is 28.6. Therefore, Johnson Controls is a possible good value selection. On the other hand, Johnson may have an expected return of 18% while General Electric may have an expected return over the next three years of 25%. We note that Emerson Electric and Johnson Controls are more likely value stocks while General Electric has a higher return on capital. Any one of these three stocks might be considered for selection.

Again if we look at the four firms in the computer hardware industry, we see that the best value is Compaq, but the expected return may only be 20% while we could expect 35% from Cisco or EMC. If we seek a higher return from Cisco then we must assume a higher risk since the PCF is very high and could readjust at any disappointment.

Finally, consider the four firms in the retail stores industry. The best choices are Gap or Dollar General, which could return 25% or 30% annually. Dollar General provided 31% annually over the period 1993–98 and could repeat that performance over the next five years.

In the next chapter, we assemble a comprehensive company profile that will help us select the best stocks.

**Table 7.10**  **The Power and Value Index for 12 Firms in Three Industries for the Forward Period 1999–2003**

| Name | Rcap | ×G | × Financial Strength | × Pr. Gr. Pers. / 100 | = Power | Value Line PCF | Actual Price | Estimated Cash Flow | Est. PCF | Value Index |
|---|---|---|---|---|---|---|---|---|---|---|
| General Electric | 26.25 | 13.5 | 1.40 | 1.0 | 496.1 | 16.0 | 95 | 3.85 | 24.7 | 20.1 |
| Emerson Electric | 21.25 | 11.0 | 1.40 | 1.0 | 327.3 | 11.5 | 63 | 4.55 | 13.85 | 23.6 |
| Johnson Controls | 12.75 | 13.0 | 1.20 | 0.95 | 189.0 | 5.5 | 64 | 9.70 | 6.60 | 28.6 |
| Honeywell | 15.25 | 10.5 | 1.20 | 0.80 | 153.7 | 9.5 | 75 | 7.70 | 9.74 | 15.8 |
| Cisco | 28.0 | 28.0 | 1.40 | 1.00 | 1097.6 | 25.0 | 96 | 1.8 | 53.3 | 20.6 |
| EMC | 22.75 | 30.0 | 1.30 | 0.95 | 842.9 | 30.0 | 92 | 2.55 | 36.1 | 23.3 |
| Compaq | 19.25 | 25.0 | 1.40 | 0.80 | 539.0 | 12.5 | 46 | 2.25 | 20.4 | 26.4 |
| Sun Microsys. | 22.25 | 16.0 | 1.30 | 0.75 | 351.0 | 13.0 | 97 | 4.05 | 23.9 | 14.7 |
| Gap | 38.5 | 26.0 | 1.30 | 0.70 | 910.9 | 14.0 | 62 | 2.35 | 26.4 | 34.5 |
| Dollar General | 23.0 | 20.5 | 1.20 | 0.95 | 537.5 | 15.0 | 25.0 | 1.55 | 16.1 | 33.4 |
| Wal-Mart | 19.25 | 16.5 | 1.30 | 0.45 | 185.8 | 16.0 | 81 | 3.35 | 24.2 | 7.7 |
| Claire's Stores | 18.75 | 15.5 | 1.10 | 0.65 | 207.8 | 10.0 | 19 | 2.15 | 8.8 | 23.6 |

## Intrinsic Value

The **Intrinsic Value** (IV) of a stock is the real or warranted value of a stock. We may calculate IV by several methods as illustrated in this chapter. The most conservative method is to use the Value Line price to cash flow (PCF) long-term correlation number as provided on the Value Line page for each company. The less conservative method is to use the PCF calculated and recorded in Table 7.1 for the estimated growth of cash flow and the estimated discount rate.

Consider an example for a company, ABC Corp. It has an estimated 2000 cash flow of $1.50 and a Value Line PCF of 13.0. Then, as shown in Table 7.11, the intrinsic value of the stock in 2000 is $19.50.

If we assume that the discount rate in 2000 is $r = 0.10$ and the expected growth of cash flow for ABC Corp. is $g = 0.09$ (9%), then we use Table 7.1 to determine that PCF = 14.42. Therefore, the intrinsic value of the stock is

$$IV = PCF \times CF$$
$$= 14.42 \times 1.50 = \$21.63.$$

The intrinsic value using these two methods of calculation is recorded in Table 7.11.

The actual price of ABC Corp. will vary from its intrinsic value as shown in Figure 7.3. The price will swing above and below its intrinsic value as market information becomes available and investors reevaluate their expectations for ABC Corp. during the year 2000.

**Table 7.11  The Calculation of the Intrinsic Value of ABC Corp. for the Year 2000.**

| Method | | PCF | Cash Flow 2000 | Intrinsic Value |
|--------|------|------|------|------|
| 1. | Value Line PCF | 13.0 | $1.50 | $19.50 |
| 2. | Calculated PCF—Table 7.1 | 14.42 | 1.50 | 21.63 |

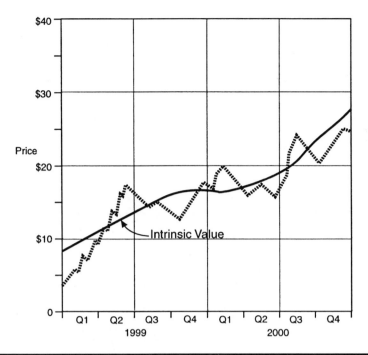

**Figure 7.3    The Intrinsic Value and the Actual Price of ABC Corp.**

# 8 | Guide to Stock Selection

## This chapter covers

- Industrial and Service Companies
- The Retail Store Industry
- The Diversified Company Industry
- The Pharmaceutical Industry
- Sales Per Employee and Cash Flow Per Employee

## The Comprehensive Selection Guide for Industrial and Service Companies

In this chapter, we will develop a stock selection guide using most of the stock performance measures we have developed in the previous chapters. However, in this chapter we only address industrial companies and service companies. These companies generate revenues using assets including net property and plant and inventory. Exhibit 8.1 provides a comprehensive analysis tool utilizing all the measurement tools of Chapters 1–7. We call this tool, the Company Selection Guide for Industrial and Service Companies.

At the top of the page enter the name, symbol and industry for the company under consideration as well as the date of the record. On Line 1 we describe the company in terms of its sales or revenues, market capitalization, total assets, net property and plant, equity capital, number of employees, inventory, and expected growth rate of sales.

The second line determines the power of the company by providing the power equation and lines to enter each element of the equation. This enables the investor to readily determine the power.

Line 3 is concerned with the price to cash flow as provided by the Value Line regression line. Entering the current price and the estimated cash flow for the forward period, one can calculate the estimated (forward) PCF. Also using the estimated PCF and the percent growth rate of cash flow will enable the investor to calculate PCFG.

Line 4 provides the calculation of the value index which is the power divided by the estimated PCF. A value index of 15 to 20 is good while a value index of 20 to 30 is excellent. A value index greater than 30 is the best, and is achieved by very few stocks.

**Exhibit 8-1.** Company Selection Guide for Industrial and Service Companies

Name:_____   Symbol: _____   Industry: _____   Date: _____

### 1. Size

| Sales ($ Mill) | Market Capitalization ($ Mill) | Total Assets ($ Mill) | Net Property and Plant ($ Mill) | Equity ($ Mill) | Number of Employees | Inventory ($ Mill) | Expected Growth Rates of Sales (%) |
|---|---|---|---|---|---|---|---|
| _____ | _____ | _____ | _____ | _____ | _____ | _____ | _____ |

### 2. Power

| Return on Capital | | G = Growth Rate of Cash Flow | | Financial Strength Rating* | | Price Growth Persistence/100 | | Power |
|---|---|---|---|---|---|---|---|---|
| _____ | x | _____ | x | _____ | x | _____ | = | _____ |

### 3. Price/Cash Flow

| Value Line PCF | Current Price | Estimated Forward Cash Flow | Estimated PCF | PCFG = Est. PCF/G |
|---|---|---|---|---|
| _____ | _____ | _____ | _____ | |

### 4. Value Index** = Power/Estimated PCF = _____ / _____ = _____

### 5. Leverage:

| % debt/Capital | Sales/Inventory | Assets/Equity | Sales/Assets |
|---|---|---|---|
| _____ | _____ | _____ | _____ |

### 6. Intellectual Capital

| $\frac{R\&D}{Sales}$ | $\frac{Sales}{Net\ Property}$ | $\frac{Sales}{Employee}$ | Foreign Sales (%) | Profit Index |
|---|---|---|---|---|
| _____ | _____ | _____ | _____ | _____ |

### 7. Return

| Profit | Operating Margin | Return on Total Assets | Return on Sales | Return on Capital | Cash Flow Employee |
|---|---|---|---|---|---|
| _____ | _____ | _____ | _____ | _____ | _____ |

### 8. Return on Equity = Return on Sales x $\frac{Sales}{Assets}$ x $\frac{Assets}{Equity}$

= _____ x _____ x _____ = _____

|  | 1 Year | 3 Years | 5 Years |
|---|---|---|---|
| 9. Actual Past Return (%) | _____ | _____ | _____ |
| Compound Annual Return (%) | _____ | _____ | _____ |
| 10. Expected Return (%) | _____ | _____ | _____ |

11. Return Confidence = Power/4 = _____  High: Greater than 100;  Medium: 50-100;  Low: Less than 50.

12. Safety and Risk:

Beta: _____   Stock Price Stability: _____   Value Line Safety Rank: _____   Est. Yield Next 12-Month: _____

---

\*   Financial Strength Rating: B=0.90, B⁺=1.0, B⁺⁺=1.10, A=1.20, A⁺=1.30, A⁺⁺=1.40.
\*\*  Value Index: Good: 10-20, Excellent: 20-30, Best: 30 or greater.

The fifth line provides four measures of leverage. The investor records the debt/capital, sales/inventory, assets/equity, and sales/assets.

Line 6 records six measures of intellectual capital. The investor records the R&D/sales, sales/net property and plant, sales/employee, percent foreign sales, and the profit index. All these measures are described in Chapter 2.

Measures of return are listed on Line 7 and return on equity is calculated on Line 8. On Line 9 we list the one-year, three-year, and five-year total return as obtained from the Value Line Survey. Then we calculate the compound annual return to yield the recorded total return.

On Line 10 we list the expected return estimated by the investor. Line 11 provides an opportunity to calculate the confidence we have in the expected return.

Finally, on Line 12 we List 4 indicators of safety or risk for the company.

## The Retail Store Industry

The discount chains continue to lead department stores in terms of growth in same store sales and number of stores. These discount retailers, such as Target and Wal-Mart, continue to take market share away from department stores. In 1997 discount store sales were $167 billion, three times than the sales of conventional department stores. They offer good prices, good value and often better quality products. They maintain their leadership by good execution, attractive prices and improving quality of products. Our approach is to compare companies in the same sub-industry as listed by Value Line. Here we focus on the retail store-discount companies and consider five successful companies:

Costco
Dayton-Hudson (Target)
Dollar General
Kohl's
Wal-Mart

Costco operates 300 wholesale-membership warehouses and is adding 25 new stores annually. The average annual return over the period 1993–98 was 36%.

Dayton-Hudson operates Target, Mervyn's, Hudson's, and Marshall Field's department stores. They plan to open 65 Target stores. Dayton-Hudson provided a 44.7% annual return over the five-year period 1993–98.

Dollar General operates discount stores in smaller shopping centers and uses smaller stores than Costco or Wal-Mart. Dollar General is a low-priced retailer of household goods and apparel. The company has about 3,700 stores, mostly in the South, Midwestern, and Mid-Atlantic states. The company plans to open 450 more stores in 1999.

Kohl's is a midwestern-based discount retailer with mid-sized stores located in strip centers or smaller centers. The company has about 220 stores in Midwestern and Mid-Atlantic states and plans to open 35 new stores in 1999.

Wal-Mart is the largest discounter in North America with estimated 1999 sales of $158 billion. Wal-Mart operates about 2800 stores in the U.S. plus 825 stores in Canada and Latin America. Wal-Marts' large cash flow allows the company to fund an ambitious expansion program.

A robust U.S. economy and value conscious consumers are supporting the discount retailers.

We will use our Company Selection Guide to profile the performance of each of the five companies, but we only provide the actual guide here for Dollar General and Wal-Mart. We summarize the profiles for all the companies in Table 8.1.

Start by examining the Company Selection Guide for Wal-Mart shown in Exhibit 8.2. When Wal-Mart went public in October 1970, a foresighted investor could have

## Table 8.1 Summary of the Profiles of Five Discount Retailers

| Name | Power | Rcap (%) | G (%) | Est. PCF | Value Index | (%) Debt Capital | Return on Assets (%) | Growth Rate of Sales (%) | Expected Return (%) |
|---|---|---|---|---|---|---|---|---|---|
| Costco | 126.3 | 12.25 | 12.5 | 25.5 | 5.0 | 24 | 7.0 | 9.0 | 30 |
| Dayton-Hudson | 113.4 | 11.75 | 13.5 | 14.8 | 7.7 | 5.1 | 6.5 | 8.5 | 30 |
| Dollar General | 537.5 | 23.0 | 20.5 | 19.4 | 27.8 | 0 | 16.9 | 21.0 | 25 |
| Kohl's | 416.7 | 15.75 | 24.5 | 37.0 | 11.3 | 26 | 11.9 | 21.5 | 30 |
| Wal-Mart | 254.5 | 18.25 | 16.5 | 25.4 | 10.0 | 33 | 10.0 | 16.0 | 25 |

**Exhibit 8-2.** *COMPANY SELECTION GUIDE FOR INDUSTRIAL AND SERVICE COMPANIES*

Name: **Wal-Mart**     Symbol: **WMT**     Industry: **Discount Retailer**     Date: **March 12, 1999**

| | Sales ($ Mill) | Market Capitalization ($ Mill) | Total Assets ($ Mill) | Net Property and Plant ($ Mill) | Equity ($ Mill) | Number of Employees | Inventory ($ Mill) | Expected Growth Rates of Sales (%) |
|---|---|---|---|---|---|---|---|---|
| 1. Size | 157,700 | 180,000 | 51,239 | 30,071 | 22,300 | 875,000 | 20,620 | 16.0 |

| | Return on Capital | | G = Growth Rate of Cash Flow | | Financial Strength Rating* | | Price Growth Persistence/100 | | Power |
|---|---|---|---|---|---|---|---|---|---|
| 2. Power | 18.25 | x | 16.5 | x | 1.30 | x | .65 | = | 254.5 |

| | Value Line PCF | Actual Price | Estimated Forward Cash Flow | Estimated PCF | PCFG = Est. PCF/G |
|---|---|---|---|---|---|
| 3. Price/Cash Flow | 16.0 | 85 | 3.35 | 25.37 | 1.54 |

4. Value Index** = Power/Estimated PCF = _____ 254.5 _____ / _____ 25.4 _____ = _____ 10.0 _____

| | % debt/Capital | Sales/Inventory | Assets/Equity | Sales/Assets |
|---|---|---|---|---|
| 5. Leverage: | 33 | 7.6 | 2.30 | 3.07 |

| | R&D Sales | Sales Net Property | Sales Employee | Foreign Sales (%) | Profit Index |
|---|---|---|---|---|---|
| 6. Intellectual Capital | — % | 5.24 | 180,228 | 6 | High |

| | Profit | Operating Margin (%) | Return on Total Assets | Return on Sales (%) | Return on Capital | Cash Flow Employee |
|---|---|---|---|---|---|---|
| 7. Return | 5,100 | 5.9 | 9.95 | 3.23 | 18.25 | 8,489 |

8. Return on Equity = Return on Sales x $\frac{Sales}{Assets}$ x $\frac{Assets}{Equity}$

= _____ 3.23 _____ x _____ 3.07 _____ x _____ 2.30 _____ = _____ 22.9 _____

| | 1 Year | 3 Years | 5 Years |
|---|---|---|---|
| 9. Actual Past Return 1/99 (%) | 117.1 | 331.2 | 236.5 |
| Compound Annual Return (%) | 117.1 | 62.7 | 27.4% |
| 10. Expected Return (%) | 25 | 25 | 25 |

11. Return Confidence = Power/4 = _____ 63.6 _____     High: Greater than 100;     Medium: 50-100;     Low: Less than 50.

12. Safety and Risk:

Beta: _1.00_     Stock Price Stability: _65_     Value Line Safety Rank: _2_     Est. Yield Next 12-Month: _0.5%_

---

\* Financial Strength Rating: B=0.90, B⁺=1.0, B⁺⁺=1.10, A=1.20, A⁺=1.30, A⁺⁺=1.40.

\*\* Value Index: Good: 10-20, Excellent: 20-30, Best: 30 or greater.

purchased 100 shares for $1650. Today (March, 1999) that holding would be worth about $8.8 million. The source of growth for Wal-Mart is the opening of new stores in the U.S. and internationally. Wal-Mart is the largest discount chain in the world and has the potential to keep growing cash flow at 15% or 16% per year.

We use the Value Line Survey and the internet or the annual report to obtain the necessary data to complete the Selection Guide for Wal-Mart. Line 1 is completed using the estimates for WMT for 1999 obtained from Value Line. The data for total assets and net property and plant can be obtained from a recent quarterly report or any internet source such as Yahoo or Excite. We also record the expected growth rate of Sales on Line 1 using Value Line estimates. We complete Line 2 and calculate the power equal to 254.5. Line 3 shows that Value Line displays a regression line of PCF = 16 while the estimated forward PCF is 25.4. Thus, the PCFG (PCF/G) is equal to 1.54 indicating Wal-Mart is reasonably valued. On Line 5 we calculate the Value Index to obtain 10.0 which is at the low end of good value. Wal-Mart has a reasonable leverage (Line 5). On Line 6 we note that foreign sales are 6% of the total and we assign a profit index equal to High (see Chapter 2). The profitability measures shown on Line 7 are all good figures for a retail discount store chain. The return on equity of 22.9 percent (Line 8) is good. Over the past five years ending Jan. 1, 1999 the firm has returned 27.4% and we predict an expected return of 25% over the next three to five years. The confidence in this return (Line 11) is medium and the safety of this stock is above average (Line 12). Overall, Wal-Mart is an attractive selection and we summarize the key figures for Wal-Mart in Table 8.1.

Now, let us turn to Exhibit 8.3, the Selection Guide for a mid-cap firm, Dollar General. The firm has projected 1999 revenues of $4,000 million and 27,000 employees. Its projected return on capital is 23% (average of 1999 and for 01–03) which is very impressive. The projected growth rate of cash flow is 20.5% and the power is calculated to be 537.5. The Value Line normal PCF is 15.0 while the projected estimated PCF is 19.35. The PCFG calculation yields 0.94 which indicates the stock is slightly undervalued. The Value Index is 27.8 which is excellent. The profitability is good with a very good return on assets (projected for 1999). We project a return of 25% and the confidence in this figure is high. The safety measures are average. The actual return for the 12 months ending January 1, 1999 was only 7.6% due to capacity constraints and start-up problems of a new distribution center.

A summary of the profiles of the discount retailers is given in Table 8.1. It is rather clear that Dollar General is the most attractive with the highest power, value index, return on capital, return on assets, and a high expected growth rate of sales.

## The Diversified Company Industry

There are many well-managed, successful diversified companies. We will study four: Allied Signal, Carlisle, General Electric, and Tyco International. In Exhibit 8.4 we display the Company Selection Guide for General Electric. Here, using the Value Line Survey, we study only the industrial portion of GE, separating out the GE Capital Services, a large credit and financial services segment of GE. This allows us to compare GE with the three other industrial products services companies. Value Line provides the data for the GE industrial services and the GE annual report was used to determine the assets and inventory of the industrial services segment as listed in Exhibit 8.4.

**Exhibit 8-3.** *Company Selection Guide for Industrial and Service Companies*

Name: __Dollar General__    Symbol: __DG__    Industry: __Discount Retailer__    Date: __March 12, 1999__

| | Sales ($ Mill) | Market Capitalization ($ Mill) | Total Assets ($ Mill) | Net Property and Plant ($ Mill) | Equity ($ Mill) | Number of Employees | Inventory ($ Mill) | Expected Growth Rates of Sales (%) |
|---|---|---|---|---|---|---|---|---|
| 1. Size | 4,000 | 5,300 | 1,299 | 477.3 | 920 | 27,400 | 944.3 | 21.0 |

| | Return on Capital | | G = Growth Rate of Cash Flow | | Financial Strength Rating* | | Price Growth Persistence/100 | | Power |
|---|---|---|---|---|---|---|---|---|---|
| 2. Power | 23.0 | x | 20.5 | x | 1.20 | x | 0.95 | = | 537.5 |

| | Value Line PCF | Actual Price | Estimated Forward Cash Flow | Estimated PCF | PCFG = Est. PCF/G |
|---|---|---|---|---|---|
| 3. Price/Cash Flow | 15.0 | 30 | 1.55 | 19.35 | 0.94 |

4. Value Index** = Power/Estimated PCF = ___537.5___ / ___19.35___ = ___27.8___

| | % debt/Capital | Sales/Inventory | Assets/Equity | Sales/Assets |
|---|---|---|---|---|
| 5. Leverage: | 0 | 4.24 | 1.41 | 3.08 |

| | R&D / Sales | Sales / Net Property | Sales / Employee | Foreign Sales (%) | Profit Index |
|---|---|---|---|---|---|
| 6. Intellectual Capital | — % | 8.38 | 145,985 | — | Medium |

| | Profit | Operating Margin (%) | Return on Total Assets | Return on Sales (%) | Return on Capital | Cash Flow / Employee |
|---|---|---|---|---|---|---|
| 7. Return | 220 | 10.3 | 16.9% | 5.50 | 23.0 | 11,853 |

8. Return on Equity = Return on Sales x $\frac{Sales}{Assets}$ x $\frac{Assets}{Equity}$

   = ___5.50___ x ___3.08___ x ___1.41___ = ___24.0___

| | 1 Year | 3 Years | 5 Years |
|---|---|---|---|
| 9. Actual Past Return __1/99__ (%) | 7.6 | 211.2 | 350.2 |
| Compound Annual Return (%) | 7.6 | 46.0 | 35.1 |
| 10. Expected Return (%) | 25 | 25 | 25 |

11. Return Confidence = Power/4 = ___134___    High: Greater than 100;    Medium: 50-100;  Low: Less than 50.

12. Safety and Risk:

   Beta: __1.25__   Stock Price Stability: __25__   Value Line Safety Rank: ___3___   Est. Yield Next 12-Month: _0.4%_

---

\* Financial Strength Rating: B=0.90, B*=1.0, B**=1.10, A=1.20, A*=1.30, A**=1.40.
\*\* Value Index: Good: 10-20, Excellent: 20-30, Best: 30 or greater.

Clearly, GE is a very powerful company with a power equal to 500.9. The cash flow is outstanding at $47,500 per employee. With a high return on capital of 26.25% and a price growth persistence equal to 100, GE has sustained an annual return of 34.2% over the past five years through 1998.

Consider the summary of the profiles of the 4 diversified industrial companies provided in Table 8.2. General Electric is the leader in power, return on capital, safety (low debt to capital) and cash flow per employee. General Electric was named *Fortune* magazine's "Most Admired Company in America" in 1998. GE has provided an average annual return of 24% over the 18 years through 1998. GE has product lines that include aircraft engines, appliances, lighting, motors, locomotives, plastics, and the

**Exhibit 8-4.** *Company Selection Guide for Industrial and Service Companies*

Name: __General Electric__    Symbol: __GE__    Industry: __Diversified__    Date: __March 17, 1999__

| | Sales ($ Mill) | Market Capitalization ($ Mill) | Total Assets ($ Mill) | Net Property and Plant ($ Mill) | Equity ($ Mill) | Number of Employees | Inventory ($ Mill) | Expected Growth Rates of Sales (%) |
|---|---|---|---|---|---|---|---|---|
| 1. Size | 57,000 | 317,000 | 80,000 | 12,000 | 40,300 | 273,000 | 5,305 | 9.0 |

| | Return on Capital | | G = Growth Rate of Cash Flow | | Financial Strength Rating* | | Price Growth Persistence/100 | | Power |
|---|---|---|---|---|---|---|---|---|---|
| 2. Power | 26.25 | x | 13.5 | x | 1.40 | x | 1.0 | = | 500.9 |

| | Value Line PCF | Actual Price | Estimated Forward Cash Flow | Estimated PCF | PCFG = Est. PCF/G |
|---|---|---|---|---|---|
| 3. Price/Cash Flow | 16.0 | $99 | 3.85 | 25.7 | 1.90 |

4. Value Index** = Power/Estimated PCF = ____500.9____ / ____25.7____ = ____19.5____

| | % debt/Capital | Sales/Inventory | Assets/Equity | Sales/Assets |
|---|---|---|---|---|
| 5. Leverage: | 2 | 10.7 | 2.0 | 0.71 |

| | R&D/Sales | Sales/Net Property | Sales/Employee | Foreign Sales (%) | Profit Index |
|---|---|---|---|---|---|
| 6. Intellectual Capital | 3.4 % | 4.75 | 333,000 | 48 | High |

| | Profit | Operating Margin (%) | Return on Total Assets | Return on Sales (%) | Return on Capital | Cash Flow/Employee |
|---|---|---|---|---|---|---|
| 7. Return | 10,500 | 20.5 | 13.1 | 18.5 | 26.25 | 47,500 |

8. Return on Equity = Return on Sales x (Sales/Assets) x (Assets/Equity)

= ____18.5____ x ____0.71____ x ____2.0____ = ____26.3%____

| | 1 Year | 3 Years | 5 Years |
|---|---|---|---|
| 9. Actual Past Return 1/99 (%) | 41.5 | 199.7 | 336.0 |
| Compound Annual Return (%) | 41.5 | 44.0 | 34.2 |
| 10. Expected Return (%) | 30% | 30% | 30% |

11. Return Confidence = Power/4 = ____125.2____    High: Greater than 100;    Medium: 50-100;  Low: Less than 50.

12. Safety and Risk:

Beta: __1.20__    Stock Price Stability: ____90____    Value Line Safety Rank: ____1____    Est. Yield Next 12-Month: __1.4%__

---

* Financial Strength Rating: B=0.90, B⁺=1.0, B⁺⁺=1.10, A=1.20, A⁺=1.30, A⁺⁺=1.40.
** Value Index: Good: 10-20, Excellent: 20-30, Best: 30 or greater.

## Table 8.2 Summary of the Profiles of Four Diversified Industrial Companies

| Name | Power | Rcap (%) | G (%) | Est. PCF | Value Index | (%) Debt Capital | Return on Assets (%) | Cash Flow Employee ($000) | Expected Return (%) |
|---|---|---|---|---|---|---|---|---|---|
| Allied Signal | 374.7 | 24.5 | 11.5 | 10.7 | 35.0 | 22 | 9.64 | 30.1 | 18.0 |
| Carlisle | 301.3 | 15.5 | 18.0 | 8.0 | 37.7 | 40 | 9.78 | 21.3 | 20.0 |
| General Electric | 500.9 | 26.25 | 13.5 | 25.7 | 19.5 | 2 | 13.1 | 47.5 | 30.0 |
| Tyco Intl. | 326.7 | 18.0 | 22.0 | 18.0 | 18.2 | 43 | 31.9 | 31.5 | 30.0 |

NBC network. It ranks first or second in most of its markets. GE management and their CEO Jack Welch are one of the top management teams in the world.

General Electric's financial unit, GE Capital Services, leases equipment, issues credit cards and finances other corporations expansions, as well as mortgage loans. With the help of the financial unit, this well-diversified company can maintain its relatively high return on capital and its steady growth of cash flow.

Tyco International is very diversified having purchased 100 companies over the period 1987–98. Tyco products include values, disposable medical supplies, firesprinkler equipment, home-security systems and electronic devices. Tyco runs its myriad units on a decentralized basis with a small central executive staff. Tyco has significant debt as a percent of capital and thus has a somewhat risky financial profile. Tyco purchased AMP, manufacturer of electric interconnect systems in 1999.

Carlisle's focus on rubber, plastic and transportation related products dates back to its founding in 1917 as a manufacturer of automobile and bicycle inner tubes. Carlisle is a well-balanced company that has managed to grow revenues consistently since 1991. Carlisle grows through acquisitions and has consistently achieved earnings growth since 1991. Value Line predicts a growth of 18% per year for cash flow.

Allied Signal consists of aerospace, control and safety, turbine technologies, engines, transportation products, and plastics. This firm has a consistent record of growing cash flow and sales since 1991 and is predicted to achieve a growth of cash flow of 11.5% over the next three to five years. Allied Signal will merge with Honeywell in late 1999.

Most investors will want to own shares of General Electric and may wish to add shares of Tyco, Carlisle, or Allied to their portfolio.

## The Pharmaceutical Industry

In 1897, Felix Hoffmann created a new industry when he developed acetylsalicylic acid known as aspirin. It turned Bayer into the world's first modern drug company. Today, sales of medical drugs amount to $300 billion a year. This is a profitable industry since the drugs are protected by patents and the firms maintain pricing power. The industry is characterized by 20-year patents, a 10-year development and approval period and large research and development costs. Thus, high-priced drugs are under patent for perhaps a 10-year sales period.

The demographics of the aging population and the increasing reliance on drug-based therapy support the future profits of successful drug companies. Furthermore, the worldwide market is growing. Drug companies tend to have higher than average debt to capital ratios than are typical and thus post higher returns on equity. They can bear higher debt levels because their sales are very consistent and predictable. Our favorite drug companies include Merck, Pfizer, Schering-Plough, and Warner-Lambert, which we study here. These firms have several winning drugs, a long pipeline of development projects, a solid research base and a great record of achievement. Profits for U.S. drug makers are expected to grow at 16% a year for the next five years. Because pharmaceuticals are usually the most cost-effective, least-invasive part of the health system, the industry expects attractive sales growth rates. The retail drug sales in the U.S. were about $100 billion in 1998 while the worldwide market was $300 billion. A

**Table 8.3  Summary of the Profiles of Four Pharmaceutical Companies**

| Name | Power | Rcap (%) | G (%) | Est. PCF | Value Index | (%) Debt Capital | Return on Assets (%) | Cash Flow Employee ($000) | Expected Return (%) |
|---|---|---|---|---|---|---|---|---|---|
| Merck | 512.0 | 37.5 | 13.0 | 25.1 | 20.4 | 19.0 | 7.1 | 132.5 | 30 |
| Pfizer | 713.3 | 34.25 | 17.5 | 40.0 | 17.8 | 17.7 | 15.4 | 79.0 | 35 |
| Schering-Plough | 883.6 | 42.5 | 16.5 | 34.4 | 25.7 | 26.3 | 12.5 | 104.0 | 35 |
| Warner-Lambert | 617.4 | 28.0 | 22.5 | 27.0 | 22.9 | 17.8 | 8.2 | 49.4 | 35 |

key driver of future profits for a drug company is the introduction of new products. Our study includes four leaders in the drug industry.

Consider the summary of the four companies given in Table 8.3. This table was developed using March 1999 estimates for the 1999 year. Schering-Plough displays the best power, value index, return on capital, return on assets, and the second-best cash-flow per employee. We show the company profile for Schering-Plough in Exhibit 8.5. This company has a high safety rank and a solid expected return of 35%. With international sales of 39% and a R&D to sales ratio of 12.5% the company generates an expected return on capital of 42.5% and an achievable cash flow growth rate of 16.5%. The Value Line correlation line calls for a PCF of 19.0. The projected PCF for 1999 was a rich 34.4. Thus, the biggest risk of this company is any decline in return on capital or growth rate of cash flow. As long as pricing power is retained, the return on capital can be maintained. Thus, at this level of valuation (using PCF), most investors would be wise to build a holding in this stock over a period of time. Alternatively, the investor could consider Warner-Lambert as a more realistically valued drug stock with an estimated 1999 PCF of 27 and a value index of 22.9.

Another attractive company is Merck which launched 14 drugs and vaccines over the period 1994–98. In 1998, new products accounted for 22% of Merck's $27 billion in sales. Merck has a solid pipeline of new drugs but will have older drug patents expire each year. Merck expected a growth rate of cash flow of 13%, the lowest for the four companies. This estimate may be low, if the new drugs are more successful than expected. It has the highest cash flow per employee and has the power to develop new opportunities. Merck's return was 37.0% annually for the five-year period ending 1998.

## Sales Per Employee and Cash Flow Per Employee

A valuable measure of the efficiency of a company is sales per employee. Industrial companies often spend a majority of their expenses on salaries and wages for their employees. Thus, if they effectively deploy their employees on important tasks they should demonstrate an attractive ratio for sales generated per employee. Table 8.4 shows the sales per employee for Gillette and Schering-Plough for the nine-year period 1990–98. Notice that both companies increased the sales generated per employee over that nine-year period. As a result of that improvement we note an improvement in profitability and stock price. In general, an industrial company should show a sales per employee ratio exceeding $200,000 as an indication of efficiency.

**Exhibit 8-5.** *COMPANY SELECTION GUIDE FOR INDUSTRIAL AND SERVICE COMPANIES*

Name: ___Schering-Plough___    Symbol: __SGP__    Industry: __Drugs__    Date: __March 18, 1999__

| | Sales ($ Mill) | Market Capitalization ($ Mill) | Total Assets ($ Mill) | Net Property and Plant ($ Mill) | Equity ($ Mill) | Number of Employees | Inventory ($ Mill) | Expected Growth Rates of Sales (%) |
|---|---|---|---|---|---|---|---|---|
| 1. Size | 9,100 | 81,000 | 7,840 | 4,068 | 4,865 | 22,700 | 841 | 14.5 |

| | Return on Capital | | G = Growth Rate of Cash Flow | | Financial Strength Rating* | | Price Growth Persistence/100 | | Power |
|---|---|---|---|---|---|---|---|---|---|
| 2. Power | 42.5 | x | 16.5 | x | 1.40 | x | 0.90 | = | 883.6 |

| | Value Line PCF | Actual Price | Estimated Forward Cash Flow | Estimated PCF | PCFG = Est. PCF/G |
|---|---|---|---|---|---|
| 3. Price/Cash Flow | 19.0 | 55 | 1.60 | 34.4 | 2.08 |

4. Value Index** = Power/Estimated PCF = ___883.6___ / ___34.4___ = ___25.7___

| | % debt/Capital | Sales/Inventory | Assets/Equity | Sales/Assets |
|---|---|---|---|---|
| 5. Leverage: | 1 | 10.8 | 1.61 | 1.16 |

| | R&D/Sales | Sales/Net Property | Sales/Employee | Foreign Sales (%) | Profit Index |
|---|---|---|---|---|---|
| 6. Intellectual Capital | 12.5 % | 2.24 | 401,000 | 39 | High |

| | Profit | Operating Margin (%) | Return on Total Assets | Return on Sales (%) | Return on Capital | Cash Flow/Employee |
|---|---|---|---|---|---|---|
| 7. Return | 2,065 | 33.0 | 26.3 | 22.7 | 42.5 | 104,000 |

8. Return on Equity = Return on Sales x (Sales/Assets) x (Assets/Equity)

= ___22.7___ x ___1.16___ x ___1.61___ = ___42.4___

| | 1 Year | 3 Years | 5 Years |
|---|---|---|---|
| 9. Actual Past Return 1/99 (%) | 79.6 | 323.5 | 614.8 |
| Compound Annual Return (%) | 79.6 | 61.8 | 48.2 |
| 10. Expected Return (%) | 35% | 35% | 35% |

11. Return Confidence = Power/4 = ___220.9___   High: Greater than 100;   Medium: 50-100;   Low: Less than 50.

12. Safety and Risk:

Beta: _1.15_   Stock Price Stability: _75_   Value Line Safety Rank: _1_   Est. Yield Next 12-Month: _0.9%_

* Financial Strength Rating: B=0.90, B⁺=1.0, B⁺⁺=1.10, A=1.20, A⁺=1.30, A⁺⁺=1.40.
** Value Index: Good: 10-20, Excellent: 20-30, Best: 30 or greater.

Table 8.5 shows the sales per employee and cash flow per employee for four industrial leaders. Note that all these companies have a sales per employee exceeding $200,000. Cash flow per employee is an excellent measure of profitability. In general we seek companies with a cash flow per employee exceeding $30,000.

**Table 8.4 Sales per Employee for Two Companies**

|  | '90 | '91 | '92 | '93 | '94 | '95 | '96 | '97 | '98 |
|---|---|---|---|---|---|---|---|---|---|
| **Gillette** | | | | | | | | | |
| Sales/ | | | | | | | | | |
| Employee ($000) | 149.1 | 157.9 | 174.0 | 172.8 | 195.0 | 210.8 | 219.9 | 228.7 | 233.3 |
| Year End | | | | | | | | | |
| Stock Price ($) | 7.84 | 14.03 | 14.22 | 14.91 | 18.72 | 26.02 | 38.88 | 50.22 | 47.81 |
| **Schering-Plough** | | | | | | | | | |
| Sales/ | | | | | | | | | |
| Employee ($000) | 166.2 | 180.8 | 202.8 | 208.3 | 226.9 | 253.9 | 274.6 | 298.6 | 321.8 |
| Net Income | | | | | | | | | |
| ($ millions) | 565 | 646 | 720 | 825 | 922 | 1,053 | 1,213 | 1,444 | 1,756 |

**Table 8.5 Cash Flow per Employee and Sales per Employee for Four Companies for 1999**

| Company | Cash Flow per Employee | Sales per Employee | Average Annual Return, 1993–98 |
|---|---|---|---|
| General Electric | $47,500 | $333,000 | 34.2% |
| Merck | $132,500 | $555,000 | 37.0% |
| Schering Plough | $104,000 | $401,000 | 48.2% |
| Tyco International | $31,500 | $226,000 | 43.0% |

# 9 Banks, Insurance, and Financial Services

## This chapter covers

- Stock Selection of Banks and Thrift Companies
- Stock Selection of Insurance Companies
- Securities Brokers and Investment Management Companies
- Financial Services Companies

## Stock Selection of Banks and Thrift Companies

A bank takes deposits from individuals and corporations and lends these funds to borrowers. When banks lend money, they provide two services. They provide a matching of lenders and borrowers and they check that the borrower is likely to repay the loan. Banks profit on the spread between what they pay depositors and what they charge borrowers. Banks often focus on commercial loans. The thrift companies and savings and loans are established to primarily provide loans for housing and construction purposes.

The best conditions for banks are steep yield curves when short-term rates are low compared to long-term rates. The banks and thrifts profit from taking deposits and paying lower short-term rates while issuing loans at higher long-term rates. When the economy is strong, then banks are also strong and profitable. Furthermore, banks have merged in recent years and consolidation will probably continue for several years. Consolidation results in reduced costs, thus, increasing returns. For example, merged banks can share the cost of new computer systems as electronic banking activity grows.

The largest U.S. bank is Citigroup with $701 billion in assets. Bank of America, the next largest with $595 billions in assets (as of March 1999). These banks were active in large mergers in 1998 and are working to manage these large institutions. Several large banks are active overseas and are subject to the problems of emerging nations.

The thrift industry, as defined by Value Line, includes large Savings Banks like Washington Mutual with $170 billion in assets. This industry also includes Fannie Mae, the nations largest provider of residential mortgage funds. Its primary function is to provide a secondary market for residential mortgages. Fannie Mae has about $470 billion in mortgage loans.

We will use the Company Selection Guide for Banks and Thrift Companies shown in Exhibit 9.1. On Line 1 we record the total loans outstanding, market capitalization and assets as well as equity, number of employees and growth rate of loans. For banks and thrifts we use return on equity and growth rate of earnings along with the financial rating and price growth persistence to calculate the power on Line 2. On Line 3 we record the normal PE as given by Value Line and calculate the estimated forward PE. Since cash flow has little meaning for financial companies, we use earnings and return on equity. We calculate the PEG by calculating

$$PEG = \frac{Forward\ PE}{Growth\ Rate\ of\ Earnings}.$$

In Line 4 we calculate the Value Index and on Line 5 we record two leverage ratios. Line 6 uses income/employee and the profit index to estimate the intellectual capital.

**Exhibit 9-1.** *Company Selection Guide for Banks and Thrift Companies*

Name: _____    Symbol: _____    Date: _____

| | Loans ($ Mill) | Market Capitalization ($ Mill) | Total Assets ($ Mill) | Equity ($ Mill) | Number of Employees | Expected Growth Rate of Loans |
|---|---|---|---|---|---|---|
| 1. Size | _____ | _____ | _____ | _____ | _____ | _____ |

| | Return on Equity | G = Growth Rate of Earnings | Financial Strength Rating* | Price Growth Persistence/100 | Power |
|---|---|---|---|---|---|
| 2. Power | _____ x | _____ x | _____ x | _____ = | _____ |

| | Value Line PE | Current Price | Estimated Forward Earnings | Estimated Forward PE | PEG = PE/G |
|---|---|---|---|---|---|
| 3. Price/Earnings | _____ | _____ | _____ | _____ | _____ |

4. Value Index** = Power/Estimated PE = _____ / _____ = _____

| | % debt/Capital | Assets/Equity |
|---|---|---|
| 5. Leverage: | _____ | _____ |

| | Income Employee | Profit Index |
|---|---|---|
| 6. Intellectual Capital | _____ | _____ |

| | Profit | Return on Total Assets |
|---|---|---|
| 7. Return | _____ | _____ |

8. Return on Equity = _____

| | 1 Year | 3 Years | 5 Years |
|---|---|---|---|
| 9. Actual Past Return | _____ | _____ | _____ |
| Compound Annual Return | _____ | _____ | _____ |
| 10. Expected Return | _____ | _____ | _____ |

11. Return Confidence = Power/4 = _____    High: Greater than 100;   Medium: 50-100,   Low: Less than 50.

12. Safety and Risk:

Beta: _____   Stock Price Stability: _____   Value Line Safety Rank: _____   Est. Yield Next 12-Month: _____

---

\*   Financial Strength Rating: B=0.90, B*=1.0, B**=1.10, A=1.20, A*=1.30, A**=1.40.
\*\*   Value Index: Good: 15-20, Excellent: 20-30, Best: 30 or greater.

Lines 7 and 8 record the return on assets and return on equity. The data on Lines 9 to 12 are similar to those for industrial companies.

Exhibit 9.2 shows the company profile for US Bancorp, a medium-sized bank with $77 billion in assets, headquartered in Minneapolis. Figure 9.1 shows the Value Line page for U.S. Bancorp. The loans portfolio consists of 20% commercial loans, 40% real estate loans, and 20% consumer loans. We calculate the power of 445.0. The Value Line normal PE is 13.0. The estimated forward PE is equal to 15.2 and PEG = 0.87. The loans, total assets and return on assets are items 1, 2, and 3 respectively on Figure 9.1. Note the slight variation in the estimate of these numbers between March 22 and April 2, 1999. This company appears to be undervalued. The Value Index is 29.9 which indicates excellent value. The company is relatively highly leveraged as shown on Line 5, as are most banks and thrifts. For example, Mellon Bank and Wells Fargo Bank have

**Exhibit 9-2.** *COMPANY SELECTION GUIDE FOR BANKS AND THRIFT COMPANIES*

Name: ___US Bancorp___    Symbol: ___USB___    Date: ___March 22, 1999___

| | Loans ($ Mill) | Market Capitalization ($ Mill) | Total Assets ($ Mill) | Equity ($ Mill) | Number of Employees | Expected Growth Rate of Loans |
|---|---|---|---|---|---|---|
| 1. Size | 58,500 | 25,400 | 77,000 | 6,500 | 26,000 | 10.0 |

| | Return on Equity | G = Growth Rate of Earnings | | Financial Strength Rating* | Price Growth Persistence/100 | | Power |
|---|---|---|---|---|---|---|---|
| 2. Power | 26.0  x | 17.5  x | | 1.0  x | 1.0 | = | 455.0 |

| | Value Line PE | Current Price | Estimated Forward Earnings | Estimated Forward PE | PEG = PE/G |
|---|---|---|---|---|---|
| 3. Price/Earnings | 13.0 | 35 | 2.30 | 15.2 | 0.87 |

4. Value Index** = Power/Estimated PE = ___455___  /  ___15.2___  =  ___29.9___

| | % debt/Capital | Assets/Equity |
|---|---|---|
| 5. Leverage: | 65 | 11.8 |

| | Income Employee | Profit Index |
|---|---|---|
| 6. Intellectual Capital | 64,400 | Medium |

| | Profit | Return on Total Assets |
|---|---|---|
| 7. Return | 1,675 | 2.18 |

8. Return on Equity  =  ___26.0___

| | 1 Year | 3 Years | 5 Years |
|---|---|---|---|
| 9. Actual Past Return thru 12/98 (%) | -3.1 | 128.9 | 295.0 |
| Compound Annual Return (%) | -3.1 | 31.8 | 31.6 |
| 10. Expected Return (%) | 25 | 25 | 20 |

11. Return Confidence = Power/4 = ___114___    High: Greater than 100;  Medium: 50-100,  Low: Less than 50.

12. Safety and Risk:

Beta: ___1.20___  Stock Price Stability: ___70___   Value Line Safety Rank: ___3___   Est. Yield Next 12-Month: ___2.0%___

---

\* Financial Strength Rating: B=0.90, B⁺=1.0, B⁺⁺=1.10, A=1.20, A⁺=1.30, A⁺⁺=1.40.
\*\* Value Index: Good: 15-20, Excellent: 20-30, Best: 30 or greater.

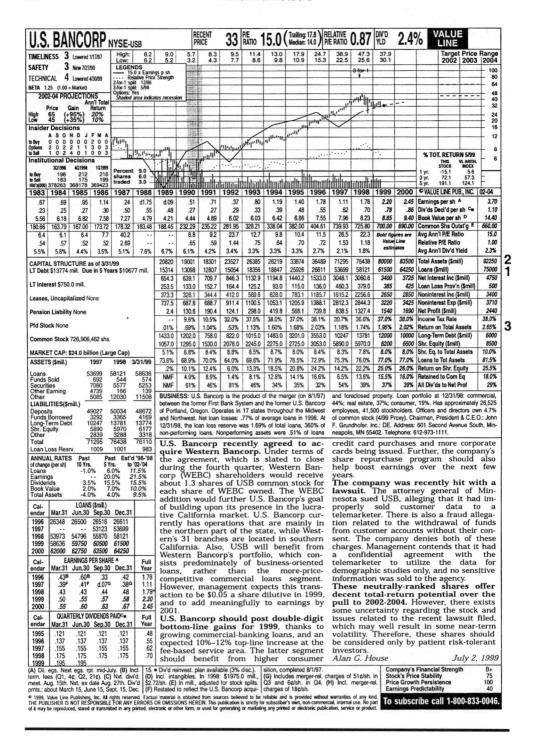

**U.S. BANCORP** NYSE-USB

| | | |
|---|---|---|
| RECENT PRICE | **33** | |
| P/E RATIO | **15.0** | (Trailing: 17.8 / Median: 14.0) |
| RELATIVE P/E RATIO | **0.87** | |
| DIV'D YLD | **2.4%** | |
| **VALUE LINE** | | |

TIMELINESS **3** Lowered 1/17/97
SAFETY **3** New 7/27/90
TECHNICAL **4** Lowered 4/30/99
BETA 1.25 (1.00 = Market)

**2002-04 PROJECTIONS**
| | Price | Gain | Ann'l Total Return |
|---|---|---|---|
| High | 65 | (+95%) | 20% |
| Low | 45 | (+35%) | 10% |

High/Low:
8.2/6.2, 9.0/5.2, 5.7/3.2, 8.3/4.3, 9.5/7.7, 11.4/8.6, 13.0/9.8, 17.9/10.9, 24.7/15.3, 38.9/22.5, 47.3/25.6, 37.9/30.1

Target Price Range 2002 2003 2004

LEGENDS
15.0 x Earnings p sh
Relative Price Strength
2-for-1 split 12/86
3-for-1 split 5/98
Options: Yes
*Shaded area indicates recession*

**Insider Decisions**
| | A | S | O | N | D | J | F | M | A |
|---|---|---|---|---|---|---|---|---|---|
| to Buy | 0 | 0 | 0 | 0 | 0 | 2 | 0 | 0 | 0 |
| Options | 2 | 0 | 2 | 2 | 1 | 1 | 3 | 0 | 3 |
| to Sell | 1 | 0 | 2 | 4 | 0 | 1 | 0 | 0 | 3 |

**Institutional Decisions**
| | 3Q1998 | 4Q1998 | 1Q1999 |
|---|---|---|---|
| to Buy | 198 | 212 | 216 |
| to Sell | 183 | 175 | 199 |
| Hld's(000) | 378263 | 368178 | 369423 |

Percent shares traded: 9.0 / 6.0 / 3.0

% TOT. RETURN 5/99
| | THIS STOCK | VL ARITH. INDEX |
|---|---|---|
| 1 yr. | -15.1 | 5.6 |
| 3 yr. | 72.1 | 57.3 |
| 5 yr. | 191.1 | 124.1 |

| | 1983 | 1984 | 1985 | 1986 | 1987 | 1988 | 1989 | 1990 | 1991 | 1992 | 1993 | 1994 | 1995 | 1996 | 1997 | 1998 | 1999 | 2000 | © VALUE LINE PUB., INC. 02-04 | |
|---|---|---|---|---|---|---|---|---|---|---|---|---|---|---|---|---|---|---|---|---|
| | .67 | .69 | .95 | 1.14 | .24 | d1.75 | .51 | .71 | .37 | .80 | 1.19 | 1.40 | 1.78 | 1.11 | 1.78 | **2.20** | **2.45** | | Earnings per sh A | 3.70 |
| | .23 | .25 | .27 | .30 | .50 | .55 | .48 | .27 | .27 | .29 | .33 | .39 | .48 | .55 | .62 | .70 | **.78** | **.86** | Div'ds Decl'd per sh C■ | 1.10 |
| | 5.56 | 6.18 | 6.82 | 7.58 | 7.27 | 4.79 | 4.21 | 4.44 | 4.89 | 6.02 | 6.03 | 6.42 | 6.86 | 7.55 | 7.96 | 8.23 | **8.85** | **9.40** | Book Value per sh D | 14.40 |
| | 180.66 | 163.79 | 167.00 | 173.72 | 178.32 | 183.48 | 188.45 | 232.29 | 235.22 | 281.95 | 328.21 | 338.04 | 382.00 | 404.61 | 739.93 | 725.80 | **700.00** | **690.00** | Common Shs Outst'g E | 660.00 |
| | 6.4 | 6.1 | 6.4 | 7.7 | 40.2 | -- | 8.8 | 9.2 | 23.7 | 12.7 | 9.8 | 10.4 | 11.5 | 26.5 | 22.3 | *Bold figures are* | | | Avg Ann'l P/E Ratio | 15.0 |
| | .54 | .57 | .52 | .52 | 2.69 | -- | .65 | .59 | 1.44 | .75 | .64 | .70 | .72 | 1.53 | 1.18 | *Value Line estimates* | | | Relative P/E Ratio | 1.00 |
| | 5.5% | 5.8% | 4.4% | 3.5% | 5.1% | 7.6% | 6.7% | 6.1% | 4.2% | 3.4% | 3.3% | 3.3% | 3.3% | 2.7% | 2.1% | 1.8% | | | Avg Ann'l Div'd Yield | 2.3% |

**CAPITAL STRUCTURE as of 3/31/99**
LT Debt $13774 mill. Due in 5 Years $10677 mill.
LT Interest $750.0 mill.
Leases, Uncapitalized None
Pension Liability None
Pfd Stock None
Common Stock 726,906,462 shs.

**MARKET CAP: $24.0 billion (Large Cap)**

| ASSETS ($mill.) | 1997 | 1998 | 3/31/99 |
|---|---|---|---|
| Loans | 53699 | 58121 | 58636 |
| Funds Sold | 692 | 544 | 574 |
| Securities | 7080 | 5577 | 5253 |
| Other Earning | 4739 | 166 | 139 |
| Other | 5085 | 12030 | 11508 |
| **LIABILITIES($mill.)** | | | |
| Deposits | 49027 | 50034 | 48672 |
| Funds Borrowed | 3292 | 3365 | 4169 |
| Long-Term Debt | 10247 | 13781 | 13774 |
| Shr. Equity | 5890 | 5970 | 6177 |
| Other | 2839 | 3288 | 3318 |
| Total | 71295 | 76438 | 76110 |
| Loan Loss Resrv. | 1009 | | 983 |

| | 20820 | 19001 | 18301 | 23527 | 26385 | 26219 | 33874 | 36489 | 71295 | 76438 | **80000** | **83500** | Total Assets ($mill) | 92250 |
| | 15314 | 13098 | 12807 | 15054 | 18356 | 18847 | 25926 | 26611 | 53699 | 58121 | **61500** | **64250** | Loans ($mill) | 75000 |
| | 654.3 | 639.1 | 709.7 | 846.3 | 1132.9 | 1194.8 | 1440.2 | 1533.0 | 3048.1 | 3060.6 | **3400** | **3725** | Net Interest Inc ($mill) | 4750 |
| | 253.5 | 133.0 | 152.7 | 164.4 | 125.2 | 93.0 | 115.0 | 136.0 | 460.3 | 379.0 | **385** | **425** | Loan Loss Prov'n ($mill) | 500 |
| | 373.3 | 326.1 | 344.4 | 412.0 | 569.6 | 628.0 | 783.1 | 1185.7 | 1615.2 | 2256.6 | **2650** | **2850** | Noninterest Inc ($mill) | 3400 |
| | 737.5 | 687.8 | 688.7 | 911.4 | 1100.5 | 1053.1 | 1205.9 | 1388.1 | 2812.3 | 2844.3 | **3220** | **3425** | Noninterest Exp ($mill) | 3710 |
| | 2.4 | 130.6 | 190.4 | 124.1 | 298.0 | 419.8 | 568.1 | 739.8 | 838.5 | 1327.4 | **1540** | **1690** | Net Profit ($mill) | 2440 |
| | -- | 9.6% | 10.5% | 32.0% | 37.5% | 38.0% | 37.0% | 38.1% | 39.7% | 36.6% | **37.0%** | **38.0%** | Income Tax Rate | 38.0% |
| | .01% | .69% | 1.04% | .53% | 1.13% | 1.60% | 1.68% | 2.03% | 1.18% | 1.74% | **1.95%** | **2.02%** | Return on Total Assets | 2.65% |
| | 1433.0 | 1202.0 | 758.0 | 822.0 | 1015.0 | 1483.0 | 3201.0 | 3553.0 | 10247 | 13781 | **12000** | **10000** | Long-Term Debt ($mill) | 6000 |
| | 1057.0 | 1295.0 | 1530.0 | 2076.0 | 2245.0 | 2725.0 | 2725.0 | 2933.0 | 5890.0 | 5970.0 | **6200** | **6500** | Shr. Equity ($mill) | 8500 |
| | 5.1% | 6.8% | 8.4% | 8.8% | 8.5% | 8.7% | 8.0% | 8.4% | 8.3% | 7.8% | **8.0%** | **8.0%** | Shr. Eq. to Total Assets | 10.0% |
| | 73.6% | 68.9% | 70.0% | 64.0% | 69.6% | 71.9% | 76.5% | 72.9% | 75.3% | 76.0% | **77.0%** | **77.0%** | Loans to Tot Assets | 81.5% |
| | .2% | 10.1% | 12.4% | 6.0% | 13.3% | 18.5% | 20.8% | 24.2% | 14.2% | 22.2% | **25.0%** | **26.0%** | Return on Shr. Equity | 25.5% |
| | NMF | 4.9% | 8.9% | 1.4% | 8.1% | 12.8% | 14.1% | 16.6% | 6.5% | 13.6% | **15.5%** | **16.0%** | Retained to Com Eq | 18.0% |
| | NMF | 61% | 46% | 81% | 46% | 34% | 35% | 32% | 54% | 39% | **37%** | **39%** | All Div'ds to Net Prof | 29% |

**BUSINESS:** U.S. Bancorp is the product of the merger (on 8/1/97) between the former First Bank System and the former U.S. Bancorp of Portland, Oregon. Operates in 17 states throughout the West and Northwest. Net loan losses: .77% of average loans in 1998. At 12/31/98, the loan loss reserve was 1.69% of total loans, 360% of non-performing loans. Nonperforming assets were .51% of loans and foreclosed property. Loan portfolio at 12/31/98: commercial, 44%; real estate, 37%; consumer, 19%. Has approximately 26,525 employees, 41,500 stockholders. Officers and directors own 4.7% of common stock (4/99 Proxy). Chairman, President & C.E.O.: John F. Grundhofer. Inc.: DE. Address: 601 Second Avenue South, Minneapolis, MN 55402. Telephone: 612-973-1111.

| ANNUAL RATES of change (per sh) | Past 10 Yrs. | Past 5 Yrs. | Est'd '96-'98 to '02-'04 |
|---|---|---|---|
| Loans | -1.0% | 6.0% | 11.5% |
| Earnings | | 20.0% | 21.5% |
| Dividends | 3.5% | 15.5% | 15.5% |
| Book Value | 2.0% | 7.0% | 10.0% |
| Total Assets | -4.0% | 4.0% | 9.5% |

| Cal-endar | LOANS ($mill.) | | | |
| | Mar.31 | Jun.30 | Sep.30 | Dec.31 |
|---|---|---|---|---|
| 1996 | 26348 | 26500 | 26516 | 26611 |
| 1997 | -- | -- | 53123 | 53699 |
| 1998 | 53973 | 54796 | 55870 | 58121 |
| 1999 | 58636 | 59750 | 60500 | 61500 |
| 2000 | 62000 | 62750 | 63500 | 64250 |

| Cal-endar | EARNINGS PER SHARE A | | | | Full Year |
| | Mar.31 | Jun.30 | Sep.30 | Dec.31 | |
|---|---|---|---|---|---|
| 1996 | .43B | .60B | .33 | .42 | 1.78 |
| 1997 | .39F | .41F | d.07G | .38G | 1.11 |
| 1998 | .43 | .43 | .44 | .48 | 1.78H |
| 1999 | .50 | .55 | .57 | .58 | 2.20 |
| 2000 | .55 | .60 | .63 | .67 | 2.45 |

| Cal-endar | QUARTERLY DIVIDENDS PAID C■ | | | | Full Year |
| | Mar.31 | Jun.30 | Sep.30 | Dec.31 | |
|---|---|---|---|---|---|
| 1995 | .121 | .121 | .121 | .121 | .48 |
| 1996 | .137 | .137 | .137 | .137 | .55 |
| 1997 | .155 | .155 | .155 | .155 | .62 |
| 1998 | .175 | .175 | .175 | .175 | .70 |
| 1999 | .195 | .195 | | | |

**U.S. Bancorp recently agreed to acquire Western Bancorp.** Under terms of the agreement, which is slated to close during the fourth quarter, Western Bancorp (WEBC) shareholders would receive about 1.3 shares of USB common stock for each share of WEBC owned. The WEBC addition would further U.S. Bancorp's goal of building upon its presence in the lucrative California market. U.S. Bancorp currently has operations that are mainly in the northern part of the state, while Western's 31 branches are located in southern California. Also, USB will benefit from Western Bancorp's portfolio, which consists predominately of business-oriented loans, rather than the more-price-competitive commercial loans segment. However, management expects this transaction to be $0.05 a share dilutive in 1999, and to add meaningfully to earnings by 2001.

**U.S. Bancorp should post double-digit bottom-line gains for 1999,** thanks to growing commercial-banking loans, and an expected 10%–12% top-line increase at the fee-based service area. The latter segment should benefit from higher consumer credit card purchases and more corporate cards being issued. Further, the company's share repurchase program should also help boost earnings over the next few years.

**The company was recently hit with a lawsuit.** The attorney general of Minnesota sued USB, alleging that it had improperly sold customer data to a telemarketer. There is also a fraud allegation related to the withdrawal of funds from customer accounts without their consent. The company denies both of these charges. Management contends that it had a confidential agreement with the telemarketer to utilize the data for demographic studies only, and no sensitive information was sold to the agency.

**These neutrally-ranked shares offer decent total-return potential over the pull to 2002-2004.** However, there exists some uncertainty regarding the stock and issues related to the recent lawsuit filed, which may well result in some near-term volatility. Therefore, these shares should be considered only by patient risk-tolerant investors.

Alan C. House          July 2, 1999

(A) Dil. egs. Next egs. rpt. mid-July. (B) Incl. term. fees (Q1, 4¢; Q2, 21¢). (C) Incl. div'd. meet. Aug. 15th. Nxt. ex date Aug. 27th. Div'd pmts.: about March 15, June 15, Sept. 15, Dec.
(D) Incl. intangibles. In 1998: $1975.0 mill., $2.72/sh. (E) In mill., adjusted for stock splits. (F) Restated to reflect the U.S. Bancorp acquisition, completed 8/1/97.
(G) Includes merger-rel. charges of 51¢/sh. in Q3 and 6¢/sh. in Q4. (H) Incl. merger-rel. charges of 18¢/sh.
15. ■ Div'd reinvest. plan available (3% disc.).

| Company's Financial Strength | B+ |
|---|---|
| Stock's Price Stability | 75 |
| Price Growth Persistence | 100 |
| Earnings Predictability | 40 |

**Figure 9.1    Value line page for US Bancorp dated April 2, 1999.**

a debt to capital ratio of 56% and 49% respectively. We may expect a return of 25% annually over the next three years for U.S. Bancorp.

A summary of the profile for three banks—Fifth Third, US Bancorp, and Mellon Bank—is provided in Table 9.1. Fifth Third, with assets of $316 billion, is a regional bank headquartered in Cincinnati primarily serving the midwest U.S. This bank has the

**Table 9.1    Summary of the Profile for Three Banks and Two Thrift Companies**

| Name | Power | RoE (%) | Growth Rate of Earnings (%) | Est PE | Value Index | Return on Assets (%) | ($) Income Employee | Percent Growth Rate of Loans | Expected Return (%) |
|---|---|---|---|---|---|---|---|---|---|
| Fifth Third Banc | 284.6 | 17.0 | 15.5 | 30.6 | 9.30 | 2.10 | 77,700 | 9.0 | 25 |
| US Bancorp | 455.0 | 26.0 | 17.5 | 15.2 | 29.9 | 2.18 | 64,400 | 10.0 | 25 |
| Mellon Bank | 408.6 | 23.0 | 17.0 | 18.6 | 22.0 | 1.89 | 32,600 | 8.0 | 25 |
| Fannie Mae | 463.5 | 23.0 | 15.5 | 19.7 | 23.5 | 0.78 | 120,300 | 12.5 | 25 |
| Washington Mut | 423.8 | 18.5 | 24.5 | 12.9 | 32.9 | 1.15 | 100,100 | 2.5 | 15 |

lowest power rating of the 3 banks and the highest PE. Thus, it has a low Value Index of 9.3. Its debt/capital ratio is 34% almost one-half of the Mellon ratio. It also has a Value Line Safety ranking of 2 (above average). This highly regarded bank has performed well over the part five years returning 38% annually.

Mellon Bank has a power of 408.6 and a Value Index of 22.0. A bank with assets of $49 billion, it is headquartered in Pittsburgh.

Considering the three banks, it may be attractive to select US Bancorp which has the best Power and Value Index. It also has a good return on assets and growth rate of loans.

The banking industry is changing with banks merging and offering broader services. The critical issues are good management, sound marketing and global presence. Banks are buying brokerages and insurance companies and consolidating the industry. These consolidations are improving efficiencies. In the future they will move more towards the ATM, the computer and the internet.

Now, we turn to the two thrift companies: Fannie Mae and Washington Mutual. Fannie Mae is a federally chartered, stockholder-owned corporation that provides secondary markets for residential loans. It is a highly leveraged company with an equity of only 4% of capital. However, it enjoys "government agency status" and has implicit backing of the U.S. government. It has a Value Line Safety rating of 2 (above average) and a financial strength of A+. It provided an average annual return of 33.5% over the five years ending December 1998. Its Power of 463.5 and Value Index of 23.5 are excellent.

Washington Mutual is the largest thrift in the U.S. with over 2,000 offices in 38 states. Washington Mutual has a Power of 423.8 and a Value Index of 32.9. This company is undervalued with a low PE. It had a negative return of −32.2% in 1998 due to problems associated with several acquisitions. The future looks good for Washington Mutual.

The safest selection is Fannie Mae while Washington Mutual may be a valuable long-term holding.

## Stock Selection of Insurance Companies

Insurance companies seek to make a profit by providing a risk-assumption service. They contract to indemnify their customers against losses arising from a peril or event that cannot be predicted at the time of the contract. One part of the insurance business, life insurance, supplements its risk assumption business by engaging in lending activities. In exchange for a payment, a premium, the insurance company compensates its customer for a loss such as fire, accident, or other peril.

Insurance companies are starting to merge and a wave of consolidation is occurring because of their relatively low market valuation.

Our analysis utilizes the Selection Guide for Insurance Companies shown in Exhibit 9.3. As noted on Line 1, premiums are the income of the insurance company. We use return on equity and growth rate of earnings to calculate the Power. For Line 6 we use premium/employee and income/employee as well as the Profit Index to indicate the intellectual capital.

**Exhibit 9-3.** *COMPANY SELECTION GUIDE FOR INSURANCE COMPANIES*

Name: _____    Symbol: _____    Date: _____

|  | Premium Income ($ Mill) | Market Capitalization ($ Mill) | Total Assets ($ Mill) | Equity ($ Mill) | Number of Employees | Expected Growth Rate of Premiums |
|---|---|---|---|---|---|---|
| 1. Size | _____ | _____ | _____ | _____ | _____ | _____ |

|  | Return on Equity | G = Growth Rate of Earnings | Financial Strength Rating* | Price Growth Persistence/100 | Power |
|---|---|---|---|---|---|
| 2. Power | _____ x | _____ x | _____ x | _____ = | _____ |

|  | Value Line PE | Current Price | Estimated Forward Earnings | Estimated Forward PE | PEG = PE/G |
|---|---|---|---|---|---|
| 3. Price/Earnings | _____ | _____ | _____ | _____ | _____ |

4. Value Index** = Power/Estimated PE = _____ / _____ = _____

|  | % debt/Capital | Assets/Equity |
|---|---|---|
| 5. Leverage: | _____ | _____ |

|  | Premium Employee | Income Employee | Profit Index |
|---|---|---|---|
| 6. Intellectual Capital | _____ | _____ | _____ |

|  | Profit | Return on Total Assets |
|---|---|---|
| 7. Return | _____ | _____ |

8. Return on Equity = _____

|  | 1 Year | 3 Years | 5 Years |
|---|---|---|---|
| 9. Actual Past Return (%) | _____ | _____ | _____ |
| Compound Annual Return (%) | _____ | _____ | _____ |
| 10. Expected Return (%) | _____ | _____ | _____ |

11. Return Confidence = Power/4 = _____    High: Greater than 100,    Medium: 50-100,    Low: Less than 50.

12. Safety and Risk:

Beta: _____  Stock Price Stability: _____  Value Line Safety Rank: _____  Est. Yield Next 12-Month: _____

---

\*　Financial Strength Rating: B=0.90, B'=1.0, B''=1.10, A=1.20, A'=1.30, A''=1.40.
\*\* Value Index: Good: 15-20, Excellent: 20-30, Best: 30 or greater.

**Table 9.2    Summary of the Profile for Five Insurance Companies**

| Name | Power | RoE (%) | Growth Rate of Earnings (%) | Est PE | Value Index | Return on Assets (%) | ($) Income Employee | Percent Growth Rate of Loans | Expected Return (%) |
|---|---|---|---|---|---|---|---|---|---|
| Aflac | 202.4 | 11.5 | 16.0 | 23.8 | 8.5 | 1.27 | 110,600 | 10.0 | 20 |
| American Intl. Gp. | 272.0 | 13.5 | 15.5 | 28.5 | 9.5 | 1.85 | 125,000 | 15.0 | 20 |
| Fremont General | 194.5 | 13.0 | 17.5 | 9.55 | 20.4 | 2.13 | 56,140 | 15.0 | 20 |
| Mercury General | 404.4 | 18.0 | 21.5 | 10.0 | 40.4 | 7.45 | 90,500 | 24.5 | 20 |
| Progressive Corp. | 272.8 | 16.0 | 15.5 | 21.5 | 12.7 | 5.13 | 34,000 | 18.5 | 20 |

We will investigate five insurance companies: Aflac, American International Group, Fremont General, Mercury General, and Progressive Corp., listed in Table 9.2. The Selection Guide for Mercury General is given in Exhibit 9.4. Mercury General is an insurance company that writes private passenger and commercial auto insurance (90% of business) and other insurance. The majority (88%) of its business is in California. It is seeking to move into other regions with its insurance products. This company operates in a very competitive industry and has experienced a slowdown in earnings growth. However, Mercury is fairly priced and the forward PE is only 10.0. The Power is 404 and the Value Index is an attractive 40.4. This is a volatile stock and it may provide a return of 20% annually over the next three years.

Alternatively, consider Fremont General which issues casualty and industrial workers compensation insurance. This company has a modest power of 194.5, but with a estimated PE of 9.55, the Value Index is 20.4. With a consistent record Fremont is a reasonable consideration but carries some risk.

The three other insurance companies with reasonable Power have a high estimated forward PE and thus a relatively low Value Index. Aflac, American International Group, and Progressive have a solid history of returns and are well regarded for their performance.

Progressive Corp. issues auto insurance in several states and has a relatively solid financial base. This company would be attractive if the price dropped 10 to 15%.

American International Group has provided solid, consistent returns over the past decade and has attained a high PE ratio. Foreign operations account for about one-half of total income. This stock has an A$^+$ financial strength and a Safety Rank of 2 (above average).

Aflac is the world's largest writer of supplemental cancer insurance. Aflac has provided an average annual return of 30% over the past 10 years thru December 1998. It is a favorite stock of many and has relatively high PE. Aflac would be attractive if its Value Index was higher as a result of a lower price (and thus lower PE).

## Securities Brokers and Investment Management Companies

Securities broker firms act as intermediaries that route securities such as stocks and bonds from the issuer or seller to the buyer. Full service brokers such as Merrill Lynch have lost some of their customers to discount brokers, such as Schwab, over the past

**Exhibit 9-4.** *COMPANY SELECTION GUIDE FOR INSURANCE COMPANIES*

Name: _____Mercury General_____     Symbol: __MCY__                    Date: __March 23, 1999__

| | Premium Income ($ Mill) | Market Capitalization ($ Mill) | Total Assets ($ Mill) | Equity ($ Mill) | Number of Employees | Expected Growth Rate of Premiums |
|---|---|---|---|---|---|---|
| 1. Size | 1,250 | 2,400 | 2,550 | 1,070 | 2,100 | 24.5 |

| | Return on Equity | | G = Growth Rate of Earnings | | Financial Strength Rating* | | Price Growth Persistence/100 | | Power |
|---|---|---|---|---|---|---|---|---|---|
| 2. Power | 18.0 | x | 21.5 | x | 1.10 | x | 0.93 | = | 404.4 |

| | Value Line PE | Current Price | Estimated Forward Earnings | Estimated Forward PE | PEG = PE/G |
|---|---|---|---|---|---|
| 3. Price/Earnings | 12.0 | 34 | 3.40 | 10.0 | 0.47 |

| | | | | | | |
|---|---|---|---|---|---|---|
| 4. Value Index** = Power/Estimated PE = | | 404.4 | / | 10 | = | 40.4 |

| | % debt/Capital | Assets/Equity |
|---|---|---|
| 5. Leverage: | 8 | 2.38 |

| | Premium Employee | Income Employee | Profit Index |
|---|---|---|---|
| 6. Intellectual Capital | 595,000 | 90,500 | Medium |

| | Profit | Return on Total Assets |
|---|---|---|
| 7. Return | 190 | 7.45% |

| | | |
|---|---|---|
| 8. Return on Equity = | 18.0 | |

| | 1 Year | 3 Years | 5 Years |
|---|---|---|---|
| 9. Actual Past Return (%) | -19.1 | 86.3 | 221.4 |
| Compound Annual Return (%) | -19.1 | 23.0 | 26.3 |
| 10. Expected Return (%) | 20 | 20 | 20 |

11. Return Confidence = Power/4 = _____101.1_____     High: Greater than 100,   Medium: 50-100,  Low: Less than 50.

12. Safety and Risk:

   Beta: _1.25_   Stock Price Stability: __50__   Value Line Safety Rank: __3__     Est. Yield Next 12-Month: _1.6%_

---

* Financial Strength Rating: B=0.90, B'=1.0, B''=1.10, A=1.20, A'=1.30, A''=1.40.
** Value Index: Good: 15-20, Excellent: 20-30, Best: 30 or greater.

decade. Recently, many brokers have instituted internet sites to allow customers to enter their orders. Schwab has been the leader of the discount brokers. It has added services for its customers over the past decade and recently added a successful internet trading business. About 85% of the trades at Schwab were on-line in 1999 versus 30% for full service brokers.

Investment management companies operate mutual funds or private investments for individuals and institutions. Mutual fund operators include T. Rowe Price and Franklin Resources. These firms make their profits by charging fees. Mutual funds hold about 20% of all retirement assets. Other related firms such as State Street handle the custodial, accounting, and recordkeeping for financial asset managers.

We use the Selection Guide for Securities Brokers and Investment Management Companies shown in Exhibit 9.5. Revenues for these companies are fees, charges, and

**Exhibit 9-5.** COMPANY SELECTION GUIDE FOR SECURITIES BROKERS AND INVESTMENT MANAGEMENT COMPANIES

Name:_____     Symbol: _____          Date: _____

| | Revenues ($ Mill) | Market Capitalization ($ Mill) | Total Assets ($ Mill) | Equity ($ Mill) | Number of Employees | Expected Growth Rate of Revenues |
|---|---|---|---|---|---|---|
| 1. Size | _____ | _____ | _____ | _____ | _____ | _____ |

| | Return on Capital | G = Growth Rate of Earnings | | Financial Strength Rating* | Price Growth Persistence/100 | Power |
|---|---|---|---|---|---|---|
| 2. Power | _____ x | _____ | x | _____ x | _____ | = _____ |

| | Value Line PE | Current Price | Estimated Forward Earnings | Estimated Forward PE | PEG = Forward PE/G |
|---|---|---|---|---|---|
| 3. Price/Earnings | _____ | _____ | _____ | _____ | _____ |

4. Value Index** = Power/Estimated PE = _____ / _____ = _____

| | % debt/Capital | Assets/Equity |
|---|---|---|
| 5. Leverage: | _____ | _____ |

| | Income/Employee | Profit Index |
|---|---|---|
| 6. Intellectual Capital | _____ | _____ |

| | Profit | Return on Total Assets | Return on Capital |
|---|---|---|---|
| 7. Return | _____ | _____ | _____ |

8. Return on Equity = _____

| | 1 Year | 3 Years | 5 Years |
|---|---|---|---|
| 9. Actual Past Return (%) | _____ | _____ | _____ |
| Compound Annual Return (%) | _____ | _____ | _____ |
| 10. Expected Return (%) | _____ | _____ | _____ |

11. Return Confidence = Power/4 = _____     High: Greater than 100,  Medium: 50-100,  Low: Less than 50.

12. Safety and Risk:

Beta:_____  Stock Price Stability: _____  Value Line Safety Rank: _____  Est. Yield Next 12-Month: _____

---

\* Financial Strength Rating: B=0.90, B'=1.0, B''=1.10, A=1.20, A'=1.30, A''=1.40.
\** Value Index: Good: 15-20, Excellent: 20-30, Best: 30 or greater.

other service relation fees. We use return on capital along with growth rate of earnings since they are reported by Value Line. We calculate Power in Line 2 and then the forward PE on line 3. The Value Index is calculated in Line 4. The remainder of the company profile is similar to other profile formats.

Exhibit 9.6 shows the Selection Guide for Charles Schwab. This firm has revenues of $3.7 billion and some 13,100 employees. With a return of capital of 18.5% and a growth rate of earnings of 17.5%, the Power is calculated to be equal to 323.8. The normal Value Line PE is 25. However, with the price at the time of this calculation, the forward PE is 75.0. Thus, the Value Index is a low 4.3. The leverage of the firm is reasonable and the Profit Index we estimate as high due to their brand awareness and innovation. We estimate an expected return of 25%. However, Schwab is overvalued at

**Exhibit 9-6.** *Company Selection Guide for Securities Brokers and Investment Management Companies*

Name: __Schwab__          Symbol: __SCH__          Date: __March 24, 1999__

| | Revenues ($ Mill) | Market Capitalization ($ Mill) | Total Assets ($ Mill) | Equity ($ Mill) | Number of Employees | Expected Growth Rate of Revenues |
|---|---|---|---|---|---|---|
| 1. Size | 3,700 | 27,300 | 18,846 | 1,750 | 13,100 | 13.0 |

| | Return on Capital | G = Growth Rate of Earnings | Financial Strength Rating* | Price Growth Persistence/100 | Power |
|---|---|---|---|---|---|
| 2. Power | 18.5  x | 17.5  x | 1.0  x | 1.00  = | 323.8 |

| | Value Line PE | Current Price | Estimated Forward Earnings | Estimated Forward PE | PEG = Forward PE/G |
|---|---|---|---|---|---|
| 3. Price/Earnings | 25 | 71 | 0.95 | 75.0 | 4.3 |

4. Value Index** = Power/Estimated PE = ___323.8___ / ___75___ = ___4.3___

| | % debt/Capital | Assets/Equity |
|---|---|---|
| 5. Leverage: | 22 | 10.8 |

| | Income Employee | Profit Index |
|---|---|---|
| 6. Intellectual Capital | $30,100 | High |

| | Profit | Return on Total Assets | Return on Capital |
|---|---|---|---|
| 7. Return | 395 | 2.10 | 18.5% |

8. Return on Equity = ___22.5%___

| | 1 Year | 3 Years | 5 Years |
|---|---|---|---|
| 9. Actual Past Return thru 12/98 (%) | 101.8 | 538.6 | 1111.1 |
| Compound Annual Return (%) | 101.8 | 85.5 | 64.7 |
| 10. Expected Return (%) | 25 | 25 | 25 |

11. Return Confidence = Power/4 = ___81___   High: Greater than 100; Medium: 50-100;   Low: Less than 50.

12. Safety and Risk:

  Beta: __1.85__   Stock Price Stability: __25__   Value Line Safety Rank: __3__   Est. Yield Next 12-Month: __0.2%__

---

\*  Financial Strength Rating:  B=0.90,  B⁺=1.0,  B⁺⁺=1.10,  A=1.20,  A⁺=1.30,  A⁺⁺=1.40.
\*\* Value Index:  Good: 15-20, Excellent: 20-30, Best: 30 or greater.

$71 unless the expected 1999 earnings are underestimated by Value Line. Several analysts expect $1.25 in earnings for 1999, and $1.60 for 2000. If the price dropped to $65 and the expected earnings are $1.25, then the forward PE is 52, still quite high. Schwab has consistently provided new innovations and attracted new customers. Perhaps it will be a good investment, if the buyer can absorb the risk inherent in the stock price and high PE ratio.

Table 9.3 provides a summary of two brokerage firms, one investment management firm and one investment management services firm. Morgan Stanley Dean Witter has an attractive valuation and provided an average annual return of 34% over the five-year period 1993–98. Morgan Stanley is realistically valued and may be a good selection for the future.

**Table 9.3    Summary of the Profiles for Four Securities Brokers, Mutual Fund Companies or Related Services**

| Name | Power | Return on Capital (%) | Growth of Earnings (%) | Est PE | Value Index | Return on Assets (%) | Debt Capital (%) | Growth Rate of Revenues (%) | Expected Return (%) |
|---|---|---|---|---|---|---|---|---|---|
| T. Rowe Price | 567.0 | 27.0 | 17.5 | 22.8 | 24.9 | 0.12 | 0 | 15.0 | 20 |
| Morgan Stanley Dean Witter | 297.6 | 16.0 | 15.5 | 18.0 | 16.5 | 0.93 | 67 | 13.5 | 20 |
| Charles Schwab | 323.8 | 18.5 | 17.5 | 75.0 | 4.3 | 2.10 | 22 | 13.0 | 25 |
| State Street | 381.8 | 18.5 | 21.5 | 26.9 | 14.2 | 1.10 | 25 | 16.5 | 20 |

T. Rowe Price provides a family of no-load funds and investment advice for private accounts. Assets under management were about $150 billion at the end of 1998. T. Rowe Price could provide a return of 20% annually for the next three years. It has a Value Index of 24.9 and is a good candidate for addition to many portfolios.

State Street is an attractively valued investment services firm. It has about $4.3 trillion under custody. This firm provided an annual return of 32% over the period 1993–98.

## Financial Services Companies

Financial services companies typically offer customers services such as credit cards, travelers checks, financial planning, and lines of credit. In this section we use the Selection Guide shown in Exhibit 9.7. This selection guide uses return on equity and growth rate of earnings to calculate Power.

The Selection Guide for MBNA is shown in Exhibit 9.8. MBNA is a lender to consumers via bank credit cards with 21 million customers. It is the leading issuer of affinity cards through membership associations. MBNA has total assets of $90 billion and earned 1.05% on them. It is relatively highly leveraged with a debt/capital ratio of 75%. The Power is 974 and its Value Index is 44.9. MBNA has provided an average annual return of 46% to its shareholders over the five years ending December 1998.

A summary of the profile of four financial service companies is provided in Table 9.4. American Express issues Travelers Checks and credit cards and provides investment management services such as mutual funds. American Express manages $121 billion in assets and earned 1.96% on these assets.

Capital One is one of the largest providers of credit cards with total assets of about $25 billion. It had a return of 3.0% on assets. It has about $14 billion in managed loans and has been a consistent performer over the three-year period 1995–98 with an average annual return of 71%.

Providian is a consumer lender through credit cards, lines of credit and home equity loans. Providian has about $18 billion in assets and a return on assets of 2.6%.

Using the Value Index as a comparative measure, MBNA, Providian, and Capital One are all attractive companies.

**Exhibit 9-7.** COMPANY SELECTION GUIDE FOR FINANCIAL SERVICES COMPANIES

Name:_____  Symbol: _____  Date: _____

| | Revenues ($ Mill) | Market Capitalization ($ Mill) | Total Assets ($ Mill) | Equity ($ Mill) | Number of Employees | Expected Growth Rate of Revenues |
|---|---|---|---|---|---|---|
| 1. Size | ____ | ____ | ____ | ____ | ____ | ____ |

2. Power

| Return on Equity | | G = Growth Rate of Earnings | | Financial Strength Rating* | | Price Growth Persistence/100 | | Power |
|---|---|---|---|---|---|---|---|---|
| ____ | x | ____ | x | ____ | x | ____ | = | ____ |

3. Price/Earnings

| Value Line PE | Current Price | Estimated Forward Earnings | Estimated Forward PE | PEG = PE/G |
|---|---|---|---|---|
| ____ | ____ | ____ | ____ | ____ |

4. Value Index** = Power/Estimated PE = _____ / _____ = _____

5. Leverage:

| % debt/Capital | Assets/Equity |
|---|---|
| ____ | ____ |

6. Intellectual Capital

| Income/Employee | Profit Index |
|---|---|
| ____ | ____ |

7. Return

| Profit ($ Mill.) | Return on Total Assets |
|---|---|
| ____ | ____ |

8. Return on Equity = _____

| | 1 Year | 3 Years | 5 Years |
|---|---|---|---|
| 9. Actual Past Return (%) | ____ | ____ | ____ |
| Compound Annual Return (%) | ____ | ____ | ____ |
| 10. Expected Return (%) | ____ | ____ | ____ |

11. Return Confidence = Power/4 = _____  High: Greater than 100;  Medium: 50-100;  Low: Less than 50.

12. Safety and Risk:

Beta: _____  Stock Price Stability: _____  Value Line Safety Rank: _____  Est. Yield Next 12-Month: _____

---

* Financial Strength Rating: B=0.90, B'=1.0, B''=1.10, A=1.20, A'=1.30, A''=1.40.
** Value Index: Good: 15-20, Excellent: 20-30, Best: 30 or greater.

## Table 9.4   Summary of the Profiles for Four Financial Services Companies

| Name | Power | Return on Equity (%) | Growth Rate of Earnings (%) | Est. PE | Value Index | Return on Assets (%) | Leverage = Assets/Equity | Growth Rate of Revenues (%) | Expected Return (%) |
|---|---|---|---|---|---|---|---|---|---|
| American Express | 218.7 | 22.5 | 13.5 | 21.1 | 10.4 | 1.96 | 11.1 | 9.0 | 15 |
| Capital One Financial | 563.8 | 20.5 | 27.5 | 26.0 | 21.7 | 3.00 | 15.6 | 15.0 | 20 |
| MBNA | 973.5 | 30.0 | 29.5 | 21.7 | 44.9 | 1.05 | 26.8 | 25.0 | 25 |
| Providian Financial | 958.5 | 30.0 | 35.5 | 33.0 | 29.0 | 2.60 | 11.7 | 20.0 | 25 |

**Exhibit 9-8.** *COMPANY SELECTION GUIDE FOR SECURITIES BROKERS AND INVESTMENT MANAGEMENT COMPANIES*

Name: __MBNA__   Symbol: __KRB__   Date: __March 24, 1999__

| | Revenues ($ Mill) | Market Capitalization ($ Mill) | Total Assets ($ Mill) | Equity ($ Mill) | Number of Employees | Expected Growth Rate of Revenues |
|---|---|---|---|---|---|---|
| 1. Size | 70,000 | 18,800 | 90,000 | 3,360 | 20,000 | 25.5 |

| | Return on Equity | G = Growth Rate of Earnings | Financial Strength Rating* | Price Growth Persistence/100 | Power |
|---|---|---|---|---|---|
| 2. Power | 30.0  x | 29.5  x | 1.10  x | 1.0  = | 973.5 |

| | Value Line PE | Current Price | Estimated Forward Earnings | Estimated Forward PE | PEG = PE/G |
|---|---|---|---|---|---|
| 3. Price/Earnings | 17.5 | 25 | 1.15 | 21.7 | 0.74 |

4. Value Index** = Power/Estimated PE = __973.5__ / __21.7__ = __44.9__

| | % debt/Capital | Assets/Equity |
|---|---|---|
| 5. Leverage: | 75 | 26.8 |

| | Income Employee | Profit Index |
|---|---|---|
| 6. Intellectual Capital | $47,000 | High |

| | Profit ($ Mill.) | Return on Total Assets |
|---|---|---|
| 7. Return | 940 | 1.05 |

8. Return on Equity = __30.0%__

| | 1 Year | 3 Years | 5 Years |
|---|---|---|---|
| 9. Actual Past Return thru 12/98 (%) | 37.3 | 266.9 | 582.6 |
| Compound Annual Return (%) | 37.3 | 54.2 | 46.8 |
| 10. Expected Return (%) | 25 | 25 | 25 |

11. Return Confidence = Power/4 = __243__   High: Greater than 100;  Medium: 50-100;  Low: Less than 50.

12. Safety and Risk:

Beta: __1.45__   Stock Price Stability: __35__   Value Line Safety Rank: __3__   Est. Yield Next 12-Month: __1.1%__

---

\* Financial Strength Rating: B=0.90, B⁺=1.0, B⁺⁺=1.10, A=1.20, A⁺=1.30, A⁺⁺=1.40.
\** Value Index: Good: 15-20, Excellent: 20-30, Best: 30 or greater.

# 10 Tortoises: Utilities, REITs, and Income Stocks

## This chapter covers

- The Tortoise Portfolio
- Utilities
- Real Estate Stocks
- Income Stocks

## The Tortoise Portfolio

The Tortoise subportfolio is structured to include the stocks of companies that grow slowly and steadily and offer a relatively secure return. Companies in a Tortoise portfolio typically have a growth rate of sales less than 8% and a payout ratio for dividends in the range of 50% to 80% of earnings. The stocks of these companies typically provide about 10% to 12% as an annual total return. They are companies with fewer growth opportunities and a steady market for their products or services. The yield of the typical tortoise stock is about twice the yield of the average stock. The yield of the S&P 500 stocks in mid-1999 was about 1.4%. Thus, tortoises usually yield about 3% or greater. Also, the annual growth rate of these dividends is approximately 3%. If a tortoise stock can return 12% annually, then we can expect a doubling of the value of our holding in six years, if we reinvest the dividends. **Total return** is the sum of a stock's appreciation plus reinvested dividends. During the period 1945–98, reinvested dividends accounted for 33% of the total return. Reinvesting dividends is a great way to add to long-term returns. Doing so can be especially profitable during market downturns, since each dollar is buying more stock as the price declines. Lots of companies offer dividend-reinvestment plans that make this easy.

These Tortoise stocks are also quite safe. They have a Value Line safety rating of 1 or 2 and a price stability index of 85% or greater.

The stocks we consider for this portfolio are usually utilities, REITs or other companies with a yield greater than 3% and a high safety rating. The primary reasons for holding dividend-paying stocks is to provide some cash payments so that you do not need to sell stocks in order to raise cash and to provide a safe subportfolio to anchor

**Table 10.1    The Percent Correlation Between Categories of Stock Portfolios**

|  | Large Growth | Large Value | Natur. Resour. | Tech. | Utility | Financial | Real Estate | Foreign Stock |
|---|---|---|---|---|---|---|---|---|
| Large Growth | 100 | 83 | 32 | 81 | 52 | 68 | 22 | 54 |
| Large Value |  | 100 | 50 | 55 | 61 | 87 | 37 | 64 |
| Natural Resources |  |  | 100 | 29 | 24 | 39 | 39 | 28 |
| Technology |  |  |  | 100 | 26 | 37 | −13 | 36 |
| Utilities |  |  |  |  | 100 | 52 | 35 | 33 |
| Financial |  |  |  |  |  | 100 | 38 | 55 |
| Real Estate |  |  |  |  |  |  | 100 | 25 |
| Foreign Stock |  |  |  |  |  |  |  | 100 |

**Source:** MorningStar.

the overall portfolio in times of price declines and thus avoiding panic during stressful times.

The percent correlation between categories of stock portfolios is shown in Table 10.1. Note that the correlation of the large capitalization growth stocks with the utilities and the real estate portfolios is only 52% and 22% respectively. Thus, these Tortoise stocks provide significant diversification in contrast to the characteristics of the Hare and Gazelle portfolios.

## Utilities

A **utility** is a public service company that provides electric, gas, or water to its customers. Utilities are usually excellent defensive stocks. The electric utility industry is moving towards deregulation with a split towards three subsectors: 1) power generation, 2) transmission, and 3) distribution. In addition, a number of electric power companies have merged with gas distributors. The electric power generators compete in an increasingly deregulated market and retain the most risk. Deregulation is expected to result in increased innovation and greater efficiency. A utility provides a reasonable yield and often has 50% of its capital as debt. Thus, they are sensitive to interest rates and act somewhat like bonds. That is, the utility stock prices usually increase when interest rates decline and vice versa.

We look for utilities with long-term safety, some growth of sales and dividends and good financial strength. Also, we seek companies with reasonable valuations as evidenced by a solid Value Index and relatively low price to cash flow. Companies with sound fundamentals should be able to provide an annual return of 12%.

Many investors have limited confidence in electric utility companies to grow earnings through mergers and through new units that build power plants around the world and sell electricity in fledgling wholesale and retail markets. Therefore, we need to find companies with good management that can handle the merger or acquisition process effectively.

One test of a good electric or gas utility is its ability to keep its costs and prices low. Ideally, a utility should be able to provide both gas and electricity to its customers. Examples of solid utilities are Duke Energy, Southern Company, FPL Group, New Century, and NIPSCO. Dominion Resources' planned merger with Consolidated

Natural Gas will result in a strong company with excellent future potential. Another example of an attractive planned merger is Scana Corporation (electric) and Public Service of North Carolina (gas).

Natural gas gathering and distribution companies offer natural gas to their customers. Due in part to the National Energy Policy Act of 1992 and the Clean Air Act amendments passed in 1990, gas is expected to displace oil, electricity, and coal in traditional industrial, commercial, residential, and power applications. Increased consolidation within the industry as well as a gradual convergence with the electric industry is also expected. Companies like Nicor and El Paso Energy offer natural gas which is the most cost-efficient fuel for heating homes and buildings and is the most environmentally friendly fossil fuel. Other examples of attractive natural gas companies are Enron and MCN Energy.

Water utilities collect, treat, and deliver water to their customers. Water utilities are good defensive stocks and many opportunities for growth through consolidation exist. There are fewer than two dozen publicly traded water utility companies. The largest of these is American Water Works. Investor-owned water utilities share many characteristics with investor-owned electric utilities. Water utilities are capital-intensive, are heavily regulated, and are continually affected by public concerns. They, too, are natural monopolies. For example, Philadelphia Suburban has acquired Consumers Water Company, becoming the second-largest investor-owned water utility in the U.S. The water utilities are considered to be recession-proof. Many small systems—which constitute the vast majority of water providers—lack the size, management and capital to upgrade facilities and comply with new laws. That is allowing investor-owned companies to expand through acquisitions.

We will use the Stock Selection Guide shown in Exhibit 10.1 for utility stocks. This guide incorporates the usual factors for industrial stocks but adds several factors regarding dividends. In this case, we add the expected growth rate of dividends on Line 1 and the percent payout on Line 12. In order to calculate the Power for a utility we use

$$\text{Power} = \text{RoE} \times \text{G} \times \text{Fin. Str.} \times \frac{\text{Stk. Pr. Stabil.}}{100},$$

where RoE = return on equity, Fin. Str. is the financial strength, Stk. Pr. Stabil. is the stock price stability and

$$\text{G} = (\text{growth rate of earnings or cash flow}) + \text{Yield}.$$

We use the growth rate of earnings or cash flow by selecting the larger number. Which rate is larger depends on several company factors and it is wisest to use the larger number. Thus, if a company had earnings growth of 7.7%, a cash flow growth of 8.1% and a yield of 5.2% we calculate

$$\text{G} = 8.1 + 5.2 = 13.3.$$

We utilize the price to cash flow to calculate the value index.

Consider the Stock Selection Guide for Duke Energy shown in Exhibit 10.2. The figure for the total assets was obtained from the 1998 annual report. The equity number can also be obtained from the annual report or an internet site such as Yahoo.

**Exhibit 10-1.** *COMPANY SELECTION GUIDE FOR UTILITIES.*

Name: _____  Symbol: _____  Industry: _____  Date: _____

| | Revenues ($ Mill) | Market Capitalization ($ Mill) | Total Assets ($ Mill) | Net Property and Plant ($ Mill) | Equity ($ Mill) | Number of Employees | Growth Rate of Dividends (%) | Growth Rate of Revenues (%) |
|---|---|---|---|---|---|---|---|---|
| 1. Size | _____ | _____ | _____ | _____ | _____ | _____ | _____ | _____ |

| | Return on Equity | G = Growth Rate of Earnings or CF + Yield | Financial Strength Rating* | Stock Price Stability/100 | Power |
|---|---|---|---|---|---|
| 2. Power | _____ x | _____ x | _____ x | _____ = | |

| | Last 3 Years Average PCF | Current Price | Estimated Forward Earnings | Estimated PCF | PCFG = Est. PCF/G |
|---|---|---|---|---|---|
| 3. Price/Cash Flow | _____ | _____ | _____ | _____ | _____ |

4. Value Index** = Power/Estimated PCF = _____ / _____ = _____

| | % debt/Capital | Assets/Equity | Sales/Assets |
|---|---|---|---|
| 5. Leverage: | _____ | _____ | _____ |

| | $\frac{Sales}{Employee}$ | $\frac{Sales}{Net\ Property}$ | Profit Index |
|---|---|---|---|
| 6. Intellectual Capital | _____ | _____ | _____ |

| | Profit | Return on Total Assets | Return on Sales | Return on Capital | Cash Flow Employee |
|---|---|---|---|---|---|
| 7. Return | _____ | _____ | _____ | _____ | _____ |

8. Return on Equity = Return on Sale x $\frac{Sales}{Assets}$ x $\frac{Assets}{Equity}$

= _____ x _____ x _____ = _____

| | 1 Year | 3 Years | 5 Years |
|---|---|---|---|
| 9. Actual Past Return (%) | _____ | _____ | _____ |
| Compound Annual Return (%) | _____ | _____ | _____ |
| 10. Expected Return (%) | _____ | _____ | _____ |

11. Return Confidence = Power/4 = _____  High: Greater than 50;  Medium: 30-50;  Low: Less than 30.

12. Safety and Risk:

| | Stock Price Stability: | Payout (%): | Value Line Safety Rank: | Est. Yield Next 12-Month: |
|---|---|---|---|---|
| Beta: _____ | _____ | _____ | _____ | _____ |

___

\* Financial Strength Rating: B=0.90, B'=1.0, B''=1.10, A=1.20, A'=1.30, A''=1.40.
\*\* Value Index: Good: 15-20, Excellent: 20-30, Best: 30 or greater.

Alternatively, we can use an estimate for 1999 by noting that Value Line reports that 44% is the long-term debt as a percent of capital (see Item 1 in Figure 10.1). Since the total capital is estimated by Value Line to be $17,290 million (see Item 2 in Fig. 10.1), then, the equity is .56 × 17,290 or $9,682 million. Since the growth rate of cash flow is larger than that of earnings we use

$$G = 9.5 + 3.9 = 13.4.$$

The stock price stability is 100 for Duke Energy (see Item 3 in Figure 10.1). We calculate the Power as

$$Power = 14.5 \times 13.4 \times 1.30 \times 1.0 = 252.6.$$

**Exhibit 10-2.** *COMPANY SELECTION GUIDE FOR UTILITIES.*

Name: _____Duke Energy_____ Symbol: _DUK_ Industry: _Elec. & Gas_ Date: _____April 5, 1999_____

| | Revenues ($ Mill) | Market Capitalization ($ Mill) | Total Assets ($ Mill) | Net Property and Plant ($ Mill) | Equity ($ Mill) | Number of Employees | Growth Rate of Dividends (%) | Growth Rate of Revenues (%) |
|---|---|---|---|---|---|---|---|---|
| 1. Size | 20,000 | 21,000 | 26,806 | 16,875 | 9,682 | 23,000 | 2.0 | 9.5 |

| | Return on Equity | G = Growth Rate of Earnings or CF + Yield | | Financial Strength Rating* | | Stock Price Persistence/100 | | Power |
|---|---|---|---|---|---|---|---|---|
| 2. Power | 24.5 | x | 13.4 | x | 1.30 | x | 1.0 | = | 252.6 |

| | Last 3 Years Average PCF | Current Price | Estimated Forward Earnings | Estimated PCF | PCFG = Est. PCF/G |
|---|---|---|---|---|---|
| 3. Price/Cash Flow | 8.7 | 58.0 | 6.90 | 8.40 | 0.88 |

4. Value Index** = Power/Estimated PCF = _____252.6_____ / _____8.40_____ = _____30.1_____

| | % debt/Capital | Assets/Equity | Sales/Assets |
|---|---|---|---|
| 5. Leverage: | 49.5 | 2.77 | 0.746 |

| | Sales Employee | Sales Net Property | Profit Index |
|---|---|---|---|
| 6. Intellectual Capital | 870,000 | 1.19 | High |

| | Profit | Return on Total Assets | Return on Sales | Return on Capital | Cash Flow Employee |
|---|---|---|---|---|---|
| 7. Return | 1,380 | 5.15 | 6.90 | 9.5 | $115,000 |

8. Return on Equity = Return on Sale x $\frac{Sales}{Assets}$ x $\frac{Assets}{Equity}$

= _____6.90_____ x _0.746_ x _2.77_ = _14.5_

| | 1 Year | 3 Years | 5 Years |
|---|---|---|---|
| 9. Actual Past Return (%) | 6.2 | 31.9 | 83.6 |
| Compound Annual Return (%) | 6.2 | 9.7 | 12.9 |
| 10. Expected Return (%) | 12.0 | 12.0 | 12.0 |

11. Return Confidence = Power/4 = _____63.2_____ High: Greater than 50; Medium: 30-50; Low: Less than 30.

12. Safety and Risk:

| | Stock Price | Payout | Value Line | Est. Yield |
|---|---|---|---|---|
| Beta: _0.50_ | Stability: _100_ | (%): _63_ | Safety Rank: _1_ | Next 12-Month: _3.9%_ |

* Financial Strength Rating: B=0.90, B*=1.0, B**=1.10, A=1.20, A*=1.30, A**=1.40.
** Value Index: Good: 15-20, Excellent: 20-30, Best: 30 or greater.

The estimated cash flow for 1999 is $6.90 and for a price of $58 we find the estimated PCF = 8.40. Thus, the Value Index is 30.1. This stock has a low beta, a price stability of 100 and a safety rating of 1. This is a very attractive Tortoise stock and should return about 12% annually over a five-year period.

The Power, Value Index, and other measures of performance of selected utility stocks are provided in Table 10.2. All the utility stocks described in Table 10.2 are attractive Tortoise candidates. Note that Northwestern and Philadelphia Suburban have

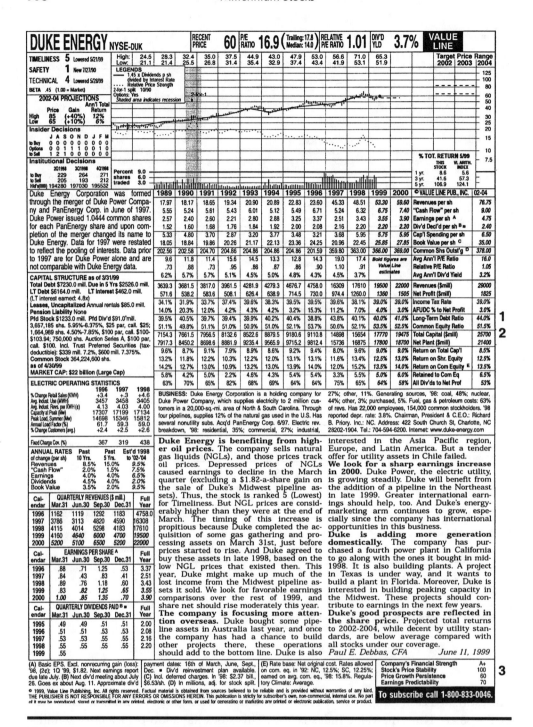

Figure 10.1   Courtesy of Value Line Survey

the highest expected growth of dividends and may be the most attractive to those seeking an increasing income stream over the next five years. If we are seeking safety as well as yield, then we look for a safety rating of 1 and a dividend greater than 4.0%. Southern Co. and New Century provide a yield of 5.4% with a safety rating of 1.

**Table 10.2 The Power and Value Index for Selected Utility Stocks**

| | G = Growth Rate of E or CF +Yield | Return on Equity (%) | Power | Value Index | Growth Rate of Sales (%) | Safety Rank | Yield (%) | Payout (%) | Growth of Dividends (%) |
|---|---|---|---|---|---|---|---|---|---|
| **Electric** | | | | | | | | | |
| Duke Energy | 13.4 | 14.5 | 252.6 | 30.1 | 9.5 | 1 | 3.9 | 63 | 2.0 |
| FPL Group | 14.0 | 12.5 | 210.0 | 48.4 | 7.5 | 2 | 4.0 | 52 | 4.0 |
| Nipsco | 11.0 | 16.5 | 217.8 | 29.4 | 16.0 | 1 | 3.5 | 61 | 6.0 |
| New Century | 13.4 | 12.5 | 201.0 | 26.5 | 4.0 | 1 | 5.4 | 71 | 2.0 |
| Northwestern | 16.5 | 11.5 | 198.3 | 25.7 | 16.0 | 2 | 4.0 | 70 | 8.0 |
| Southern | 11.4 | 11.5 | 157.3 | 29.7 | 3.5 | 1 | 5.4 | 77 | 3.5 |
| **Gas** | | | | | | | | | |
| El Paso Energy | 19.6 | 13.0 | 191.1 | 28.9 | 14.5 | 3 | 2.6 | 32 | 5.5 |
| Nicor | 10.6 | 17.5 | 241.2 | 37.4 | 6.0 | 1 | 4.1 | 55 | 4.0 |
| **Water** | | | | | | | | | |
| E 'Town | 14.9 | 10.0 | 149.0 | 18.0 | 7.5 | 2 | 4.9 | 77 | 1.5 |
| Philadelphia Suburban | 12.8 | 12.5 | 144.0 | 10.5 | 6.5 | 2 | 2.8 | 62 | 8.0 |
| Utd. Water Resources | 15.4 | 9.5 | 124.4 | 13.1 | 11.0 | 3 | 4.4 | 78 | 1.5 |

# Real Estate Stocks

Real estate business operations account for about 12% of the U.S. GDP. The real estate industry is involved in owning and operating housing, office buildings, shopping centers, and hotels. Over the past two decades, the ownership of commercial real estate has been partially transformed to publicly traded companies, primarily **Real Estate Investment Trusts** (REITs). While the majority of real estate is still privately owned, REITs account for over $200 billion in real assets. We are interested in equity REITs which are publicly traded companies. As its principal business, a REIT buys, manages, renovates, maintains and occasionally sells real properties. A REIT is not required to pay corporate taxes if it complies with several requirements, including that it pay 95% of its taxable income to shareholders in the form of cash dividends. REITs offer diversification to a portfolio as shown in Table 10.1. For example, the correlation between REITs and technology stocks is −13%. Thus, we can reduce the risk of our portfolio by adding REITs. In addition the average beta of a REIT stock is about 0.60.

A REIT has a yield of about 5% to 7% thus offering a defensive posture to market fluctuations. REITs provide a long-term average return of about 12%.

The REIT industry in the period 1994–97 experienced good growth and attracted a set of investors interested in the income growth. The investor in these years was much more short-term growth focused than long-term appreciation focused. Far less attention was paid to the building of sustainable operations for long-term shareholder value. By 1998, these investors became concerned with the potential of oversupply and 1998 was a down year for REITs. However, the cash flow performance of sound REITs continued through 1998 and into 1999. Performance should be attractive over the normal real estate cycle of six to eight years.

Real estate experiences a cycle from recession to recovery to boom to overbuilding and downtown. The cycle may average eight years in duration and the holder of REITs should be willing to absorb these cycles.

Factors to consider in evaluating a REIT include: the stock's current dividend yield; the prospects for growth in funds from operations (the REIT equivalent of net income) and thus dividends; the anticipated direction of interest rates; and the dominant property type of the company's real estate holdings and its financial fundamentals.

The performance of a REIT is measured by net income and funds from operations (FFO). FFO is the equivalent of cash flow. Thus FFO is net income excluding gains (or losses) from debt restructuring and sales of properties, plus depreciation of real property.

We use the ratio of price to FFO as the appropriate multiplier of cash flow and designate it as PFFO where

$$\text{PFFO} = \frac{\text{Price}}{\text{FFO}}.$$

The typical REIT has a PFFO of 12. (See Item 1 in Figure 10.2.) The expected growth rate for FFO over a full cycle will range from 6 to 8%. When GDP growth slows the REITs usually maintain their growth and provide a good return over an entire real estate cycle.

We will use the Selection Guide for REITs shown in Exhibit 10.3. Here we use FFO instead of cash flow. (See Item 2 in Fig. 10. 2.) In other categories the guide is similar to that used for utilities. The growth measure is

$$G = \text{Growth of FFO} + \text{Yield}$$

and the multiplier is PFFO.

The Value Line page for REITs does not provide an estimate growth rate of FFO or dividends. However, we can calculate an estimated growth rate since they give an estimated FFO and dividend per share for the current year (1999) and for three to four years forward. Using a three-year period, one can calculate the estimated growth rate. We will illustrate this with our example in Exhibit 10.6 later in this chapter.

The Selection Guide for New Plan Excel is shown in Exhibit 10.4 and the Value Line page is shown in Figure 10.2. New Plan has a Power equal to 186.3. The PFFO is calculated to be 8.4 while the average PFFO for this stock over the past three years was 12.8. Thus, this stock appears to be undervalued. Also note that the ratio of PFFO to G is 0.54 indicating undervalue. In order to calculate G, the growth rate of FFO, we note that the predicted FFO = $2.25 in 1999 and FFO = $2.75 in 2002, three years later. Thus, we can calculate the compound annual growth rate using a calculator as G = 6.9%. We also can estimate this value as the three-year percent increase divided by 3.2. Then, we have

$$\text{Percent Increase} = \frac{.50}{2.25} \times 100 = 22.2\%.$$

Therefore,

$$G = 22.2 \div 3.2 = 6.9\%.$$

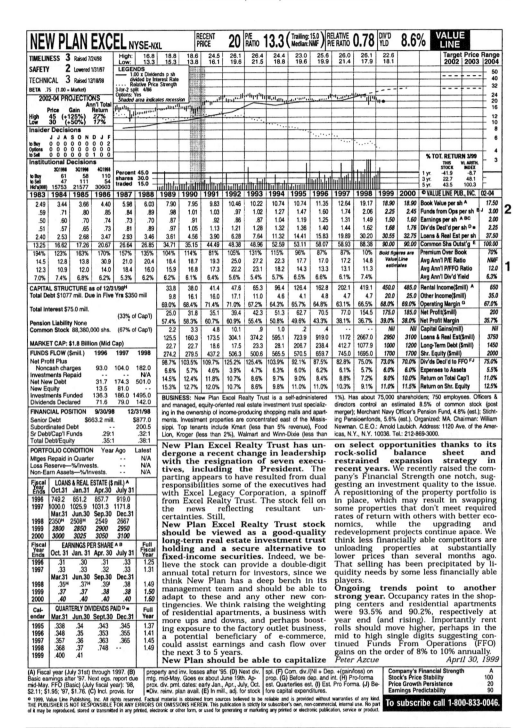

Figure 10.2    Courtesy of Value Line Publishing Inc.

The Value Index is calculated as 22.2. The debt-to-capital ratio is a low 33% and the return on equity is 11.5%. The Value Line Safety Rating is 2 and the yield is 8.5%. This is an excellent buy as a REIT in a Tortoise portfolio.

**Exhibit 10-3.** COMPANY SELECTION GUIDE FOR REITS

Name: _____     Symbol: _____              Date: _____

| | Rental Income ($ Mill) | Market Capitalization ($ Mill) | Total Assets ($ Mill) | Equity ($ Mill) | Growth Rate of Dividends (%) | Growth Rate of FFO (%) |
|---|---|---|---|---|---|---|
| 1. Size | _____ | _____ | _____ | _____ | _____ | _____ |

| | Return on Equity | G = Growth of FFO + Yield | Financial Strength Rating* | Stock Price Stability/100 | Power |
|---|---|---|---|---|---|
| 2. Power | _____ x | _____ x | _____ x | _____ = | _____ |

| | Last 3 Years Average PFFO | Current Price | Estimated Forward FFO | Estimated PFFO | PFFOG = Est. PFFO/G |
|---|---|---|---|---|---|
| 3. Price/Cash Flow | _____ | _____ | _____ | _____ | _____ |

4. Value Index** = Power/Estimated PFFO = _____ / _____ = _____

| | % debt/Capital |
|---|---|
| 5. Leverage: | _____ |

| | Profit Index |
|---|---|
| 6. Intellectual Capital | _____ |

| | Profit | Return on Total Assets |
|---|---|---|
| 7. Return | _____ | _____ |

8. Return on Equity  =  _____

| | 1 Year | 3 Years | 5 Years |
|---|---|---|---|
| 9. Actual Past Return (%) | _____ | _____ | _____ |
| Compound Annual Return (%) | _____ | _____ | _____ |
| 10. Expected Return (%) | _____ | _____ | _____ |

11. Return Confidence = Power/4 = _____   High: Greater than 40;   Medium: 20-40;   Low: Less than 30.

12. Safety and Risk:

| | Stock Price Stability: | Payout (%): | Value Line Safety Rank: | Est. Yield Next 12-Month: |
|---|---|---|---|---|
| Beta: _____ | _____ | _____ | _____ | _____ |

---

\*   Financial Strength Rating: B=0.90, B⁺=1.0, B⁺⁺=1.10, A=1.20, A⁺=1.30, A⁺⁺=1.40.
\*\*  Value Index: Good: 15-20, Excellent: 20-30, Best: 30 or greater.

Table 10.3 shows the Power and Value Index as well as other factors for selected REITs. Note the attractive valuations of these REITs. REITs provided an average return of 35% and 20% in 1996 and 1997 respectively. Since these high returns were unsustainable, the average return for REITs was −17.5% in 1998. The average annual compound return for 1996–98 was 10.2%. We can expect solid performers like Kimco and New Plan Excel to provide a return of 12% over the real estate cycle of five to seven years.

## Income Stocks

With the average dividend yields down to 1.4% in 1999, companies with safe, growing dividends are a sound defensive holding. Stocks with yields greater than 3.0% will

**Exhibit 10-4.** *COMPANY SELECTION GUIDE FOR REITS*

Name: __New Plan Excel__     Symbol: __NXL__     Date: __April 6, 1999__

|  | Rental Income ($ Mill) | Market Capitalization ($ Mill) | Total Assets ($ Mill) | Equity ($ Mill) | Growth Rate of Dividends (%) | Growth Rate of FFO (%) |
|---|---|---|---|---|---|---|
| 1. Size | 485.0 | 1,800 | 2,950 | 1,700 | 3.5 | 7.0 |

|  | Return on Equity | G = Growth of FFO + Yield | Financial Strength Rating* | Stock Price Stability/100 | Power |
|---|---|---|---|---|---|
| 2. Power | 11.5 x | 15.5 x | 1.10 x | .95 = | 186.3 |

|  | Last 3 Years Average PFFO | Current Price | Estimated Forward FFO | Estimated PFFO | PFFOG = Est. PFFO/G |
|---|---|---|---|---|---|
| 3. Price/Cash Flow | 12.8 | 19 | 2.25 | 8.4 | 0.54 |

4. Value Index** = Power/Estimated PFFO = ___186.3___ / ___8.4___ = ___22.2___

|  | % debt/Capital |
|---|---|
| 5. Leverage: | 33 |

|  | Profit Index |
|---|---|
| 6. Intellectual Capital | High |

|  | Profit | Return on Total Assets |
|---|---|---|
| 7. Return | 96.0 | 10.0 |

8. Return on Equity = ___11.5___

|  | 1 Year | 3 Years | 5 Years |
|---|---|---|---|
| 9. Actual Past Return (%) | -24.5 | 35.2 | 81.8 |
| Compound Annual Return (%) | -24.5 | 10.6 | 12.7 |
| 10. Expected Return (%) | 12.0 | 12.0 | 12.0 |

11. Return Confidence = Power/4 = ___46.6___     High: Greater than 40;   Medium: 20-40;   Low: Less than 20.

12. Safety and Risk:

|  | Stock Price |  | Payout |  | Value Line |  | Est. Yield |  |
|---|---|---|---|---|---|---|---|---|
| Beta: __0.75__ | Stability: __95__ | | (%): __72.5__ | | Safety Rank: __2__ | | Next 12-Month: __8.5%__ | |

---

\* Financial Strength Rating: B=0.90, B*=1.0, B**=1.10, A=1.20, A*=1.30, A**=1.40.
\** Value Index: Good: 15-20, Excellent: 20-30, Best: 30 or greater.

## Table 10.3   The Power and Value Index for Selected REITs

| Name | Industry | Growth of FFO +Yield | Return on Equity (%) | Power | Value Index | Growth Rate of FFO (%) | Safety Rank | Payout (%) | Yield (%) |
|---|---|---|---|---|---|---|---|---|---|
| Archstone | Apartments | 13.1 | 7.0 | 90.8 | 9.4 | 5.6 | 2 | 77.0 | 7.5 |
| BRE | Apartments | 15.3 | 9.0 | 143.9 | 15.0 | 8.2 | 2 | 59.5 | 7.1 |
| Duke | Industrial | 16.3 | 10.0 | 179.3 | 17.1 | 10.0 | — | — | 6.3 |
| Equity Res. | Apartments | 14.7 | 6.5 | 81.2 | 8.73 | 7.8 | 3 | 66.0 | 6.9 |
| Kimco | Shopping Centers | 14.4 | 8.5 | 165.6 | 16.1 | 8.2 | 2 | 63.0 | 6.2 |
| New Plan | Shopping Centers and Apartments | 15.5 | 11.5 | 186.3 | 22.2 | 7.0 | 2 | 72.5 | 8.5 |
| Rouse | Retail & Office | 13.0 | 17.5 | 184.3 | 24.3 | 7.5 | 3 | 86.0 | 5.5 |
| Weingarten | Shopping Centers | 13.6 | 13.0 | 194.5 | 19.0 | 6.5 | 2 | 72.0 | 7.1 |

hold their value in down markets. We seek companies with a good Power and Value Index as well as a Value Line safety rank of 1 or 2 and a payout less than 85%. Also, we look for solid predicted dividend growth and a proven record of providing about 12% average annual returns over a 10-year period.

We use the Selection Guide for Income Stocks as shown in Exhibit 10.5. Since we seek safety and stability, we use Stock Price Stability in the power calculation. The Selection Guide for Genuine Parts Company is given in Exhibit 10.6. Genuine Parts is a distributor of automotive replacement parts. Note that GPC has a low debt/capital and a solid return on equity. With a top safety rank of 1, this is a low-risk investment.

**Exhibit 10-5.** *Company Selection Guide for Income Stocks*

Name: _____ Symbol: _____ Industry: _____ Date: _____

| | Sales ($ Mill) | Market Capitalization ($ Mill) | Total Assets ($ Mill) | Net Property and Plant ($ Mill) | Equity ($ Mill) | Number of Employees | Growth Rate of Dividends (%) | Growth Rate of Sales (%) |
|---|---|---|---|---|---|---|---|---|
| 1. Size | ___ | ___ | ___ | ___ | ___ | ___ | ___ | ___ |

| | Return on Capital | | G = Growth Rate of Cash Flow | | Financial Strength Rating* | | Stock Price Stability/100 | | Power |
|---|---|---|---|---|---|---|---|---|---|
| 2. Power | ___ | x | ___ | x | ___ | x | ___ | = | ___ |

| | Value Line PCF | Current Price | Estimated Forward Cash Flow | Estimated PCF | PCFG = Est. PCF/G |
|---|---|---|---|---|---|
| 3. Price/Cash Flow | ___ | ___ | ___ | ___ | ___ |

4. Value Index** = Power/Estimated PCF = _____ / _____ = _____

| | % debt/Capital | Assets/Equity | Sales/Assets |
|---|---|---|---|
| 5. Leverage: | ___ | ___ | ___ |

| | Sales / Employee | Sales / Net Property | Profit Index |
|---|---|---|---|
| 6. Intellectual Capital | ___ | ___ | ___ |

| | Profit | Return on Total Assets | Return on Sales | Return on Capital | Cash Flow / Employee |
|---|---|---|---|---|---|
| 7. Return | ___ | ___ | ___ | ___ | ___ |

8. Return on Equity = Return on Sale x $\dfrac{Sales}{Assets}$ x $\dfrac{Assets}{Equity}$

= ___ x ___ x ___ = ___

| | 1 Year | 3 Years | 5 Years |
|---|---|---|---|
| 9. Actual Past Return (%) | ___ | ___ | ___ |
| Compound Annual Return (%) | ___ | ___ | ___ |
| 10. Expected Return (%) | ___ | ___ | ___ |

11. Return Confidence = Power/4 = _____ High: Greater than 50; Medium: 30-50; Low: Less than 30.

12. Safety and Risk:

| Beta: ___ | Stock Price Stability: ___ | Payout (%): ___ | Value Line Safety Rank: ___ | Est. Yield Next 12-Month: ___ |
|---|---|---|---|---|

---

* Financial Strength Rating: B=0.90, B⁺=1.0, B⁺⁺=1.10, A=1.20, A⁺=1.30, A⁺⁺=1.40.
** Value Index: Good: 15-20, Excellent: 20-30, Best: 30 or greater.

**Exhibit 10-6.** *COMPANY SELECTION GUIDE FOR INCOME STOCKS.*

Name: ___Genuine Parts___  Symbol: __GPC__  Industry: __Auto Parts__  Date: __April 8, 1999__

| | Sales ($ Mill) | Market Capitalization ($ Mill) | Total Assets ($ Mill) | Net Property and Plant ($ Mill) | Equity ($ Mill) | Number of Employees | Growth Rate of Dividends (%) | Growth Rate of Sales (%) |
|---|---|---|---|---|---|---|---|---|
| 1. Size | 7,825 | 5,400 | 3,600 | 404 | 2,240 | 24,200 | 9.0 | 10.5 |

| | Return on Capital | | G = Growth Rate of Cash Flow | | Financial Strength Rating* | | Stock Price Stability/100 | | Power |
|---|---|---|---|---|---|---|---|---|---|
| 2. Power | 15.5 | x | 10.0 | x | 1.40 | x | 1.0 | = | 217.0 |

| | Value Line PCF | Current Price | Estimated Forward Cash Flow | Estimated PCF | PCFG = Est. PCF/G |
|---|---|---|---|---|---|
| 3. Price/Cash Flow | 13.5 | 29 | 2.60 | 11.15 | 1.1 |

4. Value Index** = Power/Estimated PCF = ___217___ / ___11.15___ = ___19.5___

| | % debt/Capital | Assets/Equity | Sales/Assets |
|---|---|---|---|
| 5. Leverage: | 22 | 1.61 | 2.17 |

| | Sales / Employee | Sales / Net Property | Profit Index |
|---|---|---|---|
| 6. Intellectual Capital | 323,350 | 19.4 | Medium |

| | Profit | Return on Total Assets | Return on Sales | Return on Capital | Cash Flow / Employee |
|---|---|---|---|---|---|
| 7. Return | 385 | 10.7 | 4.92 | 15.5 | $19,400 |

8. Return on Equity = Return on Sale x $\frac{Sales}{Assets}$ x $\frac{Assets}{Equity}$

= ___4.92___ x ___2.17___ x ___1.61___ = ___17.2___

| | 1 Year | 3 Years | 5 Years |
|---|---|---|---|
| 9. Actual Past Return thru 12/98 (%) | 1.5 | 55.3 | 192 |
| Compound Annual Return (%) | 1.5 | 9.2 | 11.3 |
| 10. Expected Return (%) | 12.0 | 12.0 | 12.0 |

11. Return Confidence = Power/4 = ___54.3___  High: Greater than 50;  Medium: 30-50;  Low: Less than 30.

12. Safety and Risk:

| Beta: 0.80 | Stock Price Stability: ___ | Payout (%): 50 | Value Line Safety Rank: 1 | Est. Yield Next 12-Month: 3.6% |
|---|---|---|---|---|

* Financial Strength Rating: B=0.90, B⁺=1.0, B⁺⁺=1.10, A=1.20, A⁺=1.30, A⁺⁺=1.40.
** Value Index: Good: 15-20, Excellent: 20-30, Best: 30 or greater.

The Power, Value Index and other measures for 10 selected income stocks are given in Table 10.4. We selected stocks with a yield equal to or greater than 3.0% and with a safety rank of 1 or 2. Genuine Parts and Federal Signal are attractive since their Value Index is reasonable and the expected growth of dividends is 9.0% and 9.5% respectively. Nalco Chemical is also attractive since it has the highest Value Index, a safety rank of 1 and a yield of 3.8%.

**Table 10.4    The Power, Value Index and Other Measures for Selected Income Stocks**

| Name | Industry | Return on Capital (%) | Growth Rate of Cash Flow (%) | Power | Value Index | Growth of Div. (%) | Safety Rank | Payout (%) | Yield (%) |
|------|----------|------|------|-------|-------|------|------|------|------|
| Cedar Fair | Recreation | 19.0 | 7.5 | 135.4 | 13.3 | 3.0 | 2 | 73 | 5.3 |
| Deluxe | Publishing | 25.5 | 8.0 | 212.2 | 27.2 | 1.5 | 2 | 60 | 4.9 |
| First Union | Bank* | 21.5 | 14.0 | 270.9 | 20.2 | 15.0 | 2 | 47 | 3.5 |
| Federal Signal | Elec. Equipment | 15.5 | 11.0 | 173.9 | 17.8 | 9.5 | 2 | 50 | 3.8 |
| Genuine Parts | Auto Parts | 15.5 | 10.0 | 217.0 | 19.5 | 9.0 | 1 | 50 | 3.6 |
| Heinz | Food | 18.5 | 9.0 | 205.6 | 16.1 | 7.5 | 1 | 53 | 3.0 |
| Luby's | Restaurant | 12.0 | 7.0 | 87.8 | 13.7 | 4.5 | 2 | 53 | 5.0 |
| National City | Bank* | 20.0 | 12.0 | 259.2 | 17.8 | 10.0 | 2 | 47 | 3.3 |
| Nalco | Chemical | 18.0 | 9.0 | 175.0 | 29.6 | 2.5 | 1 | 47 | 3.8 |
| WD-40 | Chemical | 38.5 | 8.5 | 333.8 | 22.7 | 2.5 | 2 | 83 | 5.1 |

*Bank: Use RoE, growth of earnings, and PE.

# 11 Building the Contrarian Portfolio

## This chapter covers

- The Contrarian Portfolio
- Comparing Candidates for the Contra Portfolio

## The Contrarian Portfolio

The Contrarian (Contra) subportfolio described in Chapter 5 includes out-of-favor, turnaround companies with low price to book ratios or low price to cash flow. These companies were formerly strong, but earnings have declined significantly. The essence of contrarian investing is buying the stock of sound companies when others will not [Gallea, 1998]. Investors often overreact to events and devalue sectors or companies excessively. The contrarian takes advantage of these overreactions by picking up out-of-favor stocks that will survive the downturn. The key is to pick the survivors when they are down. In 1997, Ascend Communications dropped from a price of $80 at the beginning of the year to a low of $22 at the end of the year. Then, the year 1998 was a reverse with a start at a low of $24 rising to a price of $69 at the end of 1998. Finally, in early 1999 Lucent agreed to purchase Ascend at a price of $108. If one purchased at $24 in January 1998 and sold at $72 in January 1999, a gain factor of 3× would have been achieved in 12 months. Holding battered, out of favor stocks in a down market has advantages. Most who would sell the stock have already acted by the time a stock bottoms out. Look for stocks at their 104 week low that have experienced a disappointment and have been widely sold. An example is McDonald's stock price peaked in 1996 when earnings slowed and analysts said the market for fast food restaurants was shrinking. However, it turned around in 1997 and returned 68% in 1998.

It is difficult, if not impossible, in a dynamic economy, with constantly changing conditions, to use the past to accurately predict the future. Actual positive earnings surprises quickly affect the share prices of out-of-favor companies. Similarly, a negative earnings surprise quickly causes a favorite to become out of favor.

**Exhibit 11.1    Criteria for Contrarian Stock Purchase**

- The stock has provided reasonable annual returns over a period in the recent past.
- The stock price is down by 40% or more from its 104-week high.
- The price-to-book-value ratio has declined to 60% or less of its value at its 104-week high.
- The price to cash flow has declined to 70% or less of the Value Line normal value.
- The financial position of the stock is at least B$^+$ by Value Line and the dividend, if any, appears secure. The debt to capital ratio is relatively low.
- The cash flow predicted by Value Line for the coming year is up by at least 5% over the preceding year.
- The PCF-to-growth-of-cash-flow ratio is less than 1 for the coming year.
- The company is expected to perform well in the coming year according to the analysts. There is a catalyst for this positive development.
- Insider purchases are increasing or significant as reported by Value Line.

The earnings of Xerox peaked in 1989 and declined through 1991, drawing the share price to a low of $10 (adjusted for a later split). At that time the company paid a dividend of $1 (a yield of 10%). The earnings were essentially flat from 1990 to 1993. However, in 1994 the cash flow and earnings started up, and Xerox was again on an upward price path.

Acting to purchase out-of-favor stocks requires sound reasoning and the assumption of risk. The criteria for success in contrarian stock selection are listed in Exhibit 11.1. Consider Dell Computer in 1993, when the company reported negative earnings and cash flow. The stock price dropped to one-half of its price held in early 1993 to a low of $0.40 (adjusted for later splits). Of course, by 1999 the price had risen to $80. Consider the criteria of Exhibit 11.1—could you have chosen Dell in 1993? The stock price decline criterion was satisfied, but the price to book value decline was only 10%. The company remained financially stable and the predicted cash flow was positive. The estimated price/cash flow ratio for 1994 was 8, and the growth rate of cash flow was 40% for the next three years. The company was expected to perform well, but it was a risky selection. If you didn't buy in 1994 at $1 (price adjusted) then you could have waited to buy at $2 in 1995 and reaped a large reward. It is easy to see a good story in retrospect. Can anyone clearly see it at the time of the drop in price? It is difficult because of all the bad news.

A selling price may offer an opportunity to pick up a good stock at an undervalued price. It may take three or more years to reap the reward, but it can be rewarding nevertheless. One of the best examples of a great turnaround was IBM. The price of IBM declined steadily during the period of 1991–1993. Then, starting in late 1993, the company's fortunes turned around and the stock price started climbing. IBM provided a total return of 589.5% over the five-year period 1993–1998 for a compound annual return of 47.1%.

Several measures of performance for IBM over the period 1991–94 are provided in Table 11.1. The company's performance declined over the period 1991–93, and the stock price declined from a high of $69.9 in 1991 to a low of $20.4 in 1993. Let us use the criteria of Exhibit 11.1 to determine if IBM would have been a good Contra stock in 1993. IBM had provided good returns in the 1980s and was a financially strong company in 1993. The stock price had declined from a high of $50.2 in 1992 to a low

**Table 11.1   Several Measures of Performance for IBM for the Period 1991–1994**

|  | 1991 | 1992 | 1993 | Predicted 1994 |
|---|---|---|---|---|
| Revenues/Share | 56.73 | 56.46 | 53.94 | 54.00 |
| Cash Flow/Share | 7.79 | 6.73 | 5.70 | 6.00 |
| Book Value/Share | 32.40 | 24.17 | 16.01 | 16.50 |
| Average Price | 52.9 | 41.8 | 25.2 | 28.6 |
| PCF | 6.8 | 6.2 | 4.4 | 4.8 |
| P/Bk | 1.63 | 1.73 | 1.57 | 1.73 |
| Earnings | 1.85 | 1.24 | −.03 | 1.30 |
| Average PE | 28.6 | 33.7 | NM | 22.0 |
| Low Price | 41.8 | 24.4 | 20.4 | — |
| PCF @ Low Price | 5.36 | 3.63 | 3.58 | — |
| P/Bk @ Low Price | 1.29 | 1.01 | 1.27 | — |
| Value Line PCF | 7.0 | 7.0 | 7.0 | — |
| High Price | 8.97 | 7.46 | 5.25 | — |
| PCF High Price | 8.97 | 7.46 | 5.25 | — |
| P/Bk @ High Price | 2.34 | 2.08 | 1.89 | — |

$20.4 in 1993. This was a decline to 41% of its high value in the 104-week period. Similarly the PCF dropped from a Value Line normal 7.40 to a low of 3.63, a drop to 49% of its normal value. The predicted cash flow was projected to increase by 5% in 1994. The PCFG ratio was 4.8/5.0 = 0.96. The main catalyst for positive development occurred in 1993 with the appointment of Louis Gerstner as Chairman and CEO. Gerstner had a proven record as president of American Express and CEO of RJR Nabisco. Thus all nine criteria were satisfied and one would have been well advised to purchase IBM for the Contra portfolio. After two years of solid performance in 1994 and 1995, the stock would have been moved to the Hare portfolio. Many commentators, in 1993, had pronounced IBM dead—but they ignored IBM's large installed base of mainframe computers that provided a steady cash flow.

Often firms make a big mistake that undermines their stock's performance. The key is to determine if they are fixing that mistake. For example, consider Toys R Us (TOY) that allowed inventories to build up. They have tried in 1998 and 1999 to counter the trends of shoppers to use discount stores as an alternative to TOY. If one examines the Value Line (1999) page, it is clear that TOY has the financial strength to work through the problems. The price to book value and price to cash flow have declined to less than one-half of the high value in 1998. Value Line projects a growth of cash flow of 5% for 1999–03 and a 14% increase in cash flow from 1998 to 1999. The inventory had grown to $3,256 million by 10/31/98 with sales of about $11,150 million in 1998. The inventory to sales ratio is abnormally high. Thus, the key event is to determine the level of inventory. We would then seek to determine this value by accessing the balance sheet for January 31, 1999. At the time of this writing, April 1999, we would not yet purchase TOY since it has not yet reduced inventory nor shown an increase in cash flow.

## Comparing Candidates for the Contra Portfolio

When choosing candidates for the Contra subportfolio, look for stocks that have recently hit a new low price over the past 52 weeks as listed in *The Wall Street Journal* or

*Investors Business Daily.* Candidates should preferably have a Financial Strength Rating as listed by Value Line of B$^+$ or better. Also, the debt-to-capital ratio should be less than 50%. This, at least, provides the companies with the financial strength to survive as a turnaround develops. Sometimes a company is out of favor due to an industry trend such as low oil prices. Others may be seen as missing or lagging an industry change such as Borders' delay in adding internet sales of its books and music CDs.

The performance measures for nine selected stocks are shown in Table 11.2. Borders lost price strength when it was seen as a laggard in meeting the challenges of Amazon.com and Barnes & Noble on the internet. The price-to-book and the price-to-cash-flow ratios have dropped significantly and yet the company's cash flow is predicted to continue to grow. The price-to-cash-flow ratio has dropped to 5.6, but the predicted growth rate, G, and the return on capital are attractive. This stock is worthy of follow-up study.

Boeing is a company with great potential for an earnings recovery. Again, the predicted growth rate, G, and the return on capital are attractive. Boeing is a management and operations turnaround. Look for performance increases to determine when to buy.

ConAgra is a slow growth food processor. A turnaround of costs will help this company; however, the potential for solid gains is limited with a modest projected growth rate of cash flow, G. Hercules is at a low for its price to cash flow, but has a low projected return on capital.

Halliburton, an oil field service company, will turn around as oil prices rise from their 1999 lows. This is a higher risk offering due to its sensitivity to oil prices, which can be quite volatile. Nucor is a minimill steel company that has experienced a five-year period, 1993–98, with flat earnings and negative average annual returns. If a reduction of steel imports is achieved, then Nucor has a potential for solid performance, but a wait and see attitude would be prudent. Pall is another company with a solid past and a good financial rating but limited immediate potential. With a revival of the semiconductor industry and a significant management response to health industry changes, this stock could shift to positive returns.

PPG, a manufacturer of glass products and specialty chemicals, was a solid performer in the period 1993–1996 but its return flattened in 1997–98. With a firming in chemical prices, the potential for growth in the period 1999–2000 is good. A return to a good growth rate would make this an attractive turnaround.

Finally, Unisys may have great potential in the competitive computer services industry. Value Line predicts a growth rate, G, of 16% and a return on capital of 17%. If this is achieved, then Unisys may be a good selection.

Examining all nine potential Contra stocks, it is clear that selecting turnaround candidates is quite difficult. A poor performer that has a good possibility of moving to only modest performance is not a great selection. We want sound companies that have a solid potential for return to good performance. Borders and Boeing have the greatest potential returns. Both have good financial ratings and solid projected growth rates and return on capital.

**Table 11.2  Performance Measures for Nine Selected Stocks in April 1999**

| Name | Industry | Price | 104-Week High | 104-Week Low | 104-Week Price/Book High | 104-Week Price/Book Now | PCF Value Line Normal | PCF Now | Financial Rating | Debt Capital (%) | Cash Flow Change 1998 to 1999 (%) | Cash Flow Predicted Growth Rate G | PCF/G | Return on Capital 1999–2000 (%) | Expected Positive Event |
|---|---|---|---|---|---|---|---|---|---|---|---|---|---|---|---|
| Borders | Book Retailer | 14 | 42 | 13 | 4.69 | 1.27 | 13.0 | 5.6 | A | 1 | 16.3 | 20.0 | 0.28 | 13.0 | Internet Sales |
| Boeing | Aircraft Manufacturer | 33 | 60 | 29 | 4.63 | 2.30 | 12.0 | 9.2 | A | 33 | 28.5 | 17.0 | 0.54 | 11.0 | Operations Improved |
| ConAgra | Food Processor | 23 | 34 | 22 | 7.06 | 3.80 | 11.0 | 9.2 | A | 46 | 6.8 | 8.5 | 1.08 | 13.5 | Meat Supply Leveling |
| Hercules | Chemicals | 24 | 54 | 24 | 7.52 | 3.04 | 11.0 | 5.3 | B | 45 | 28.6 | 8.5 | 0.63 | 9.5 | Asia, Latin America Turning |
| Halliburton | Oilfield Services | 30 | 63.3 | 25 | 6.43 | 3.2 | 14.0 | 11.1 | A | 24 | −10.0 | 13.0 | 0.85 | 11.0 | Oil Prices Rising |
| Nucor | Steel | 42 | 63 | 35 | 2.95 | 1.65 | 10.0 | 7.92 | A+ | 10 | −6.4 | 9.0 | 0.88 | 10.0 | Steel Prices Firming |
| Pall | Fluids Filters | 16 | 26 | 15 | 4.35 | 2.39 | 15.0 | 10.3 | A | 13 | 8.4 | 9.0 | 1.15 | 13.0 | Health System Adjustments |
| PPG | Glass and Chemicals | 49 | 76 | 49 | 4.70 | 2.69 | 9.0 | 7.9 | A++ | 27 | 0.0 | 8.5 | 0.93 | 17.5 | Chemical Prices Firming |
| Unisys | Computer Services | 28 | 36 | 13 | 16.3 | 8.5 | 9.0 | 10.8 | B | 48 | 20.9 | 16.0 | 0.68 | 17.0 | Recent Growth |

*Low Price in period March 15–April 15, 1999.

# 12 Global Investing

## This chapter covers

- The Global Economy
- Investing Globally
- Selecting Stocks of Foreign Companies
- Selecting U.S.-Based Transnational Companies

## The Global Economy

Globalization is one of the most powerful and pervasive influences on businesses and nations. Information technology, communication, travel and trade tie the world together in an economy that increasingly is dependent upon worldwide markets. As globalization changes the requirements for business success, what companies are best prepared to take advantages of opportunities in the global economy? What industries are best suited for international business?

The key elements of business are increasingly mobile across borders. Financial capital and intellectual capital are easily moved from country to country while human capital is mobile to a lesser extent. Ideas move quickly around the world. CNN, available in 150 countries, is viewed by at least 150 million households worldwide.

Large **transnational companies** (TNC) strive to introduce world products simultaneously in many countries. Ford strives to produce a world car while Kimberly Clark introduces world diapers. TNCs disperse across nations and increasingly look like world companies rather than a company headquartered in one place with branches elsewhere. Gillette operates everywhere and introduces new products everywhere simultaneously. Trade has made the world a smaller place. In the U.S., two-way trade has climbed from 22% of GDP in 1984 to 30% in 1998.

Global integration of business is on the way to delivering profound worldwide prosperity while causing growing pains and human costs of transition. Globalization spreads through widely available technology and the availability of information. TNCs bring advanced technology to less developed countries (LDCs) and low-cost products to wealthier countries. TNCs in 1995 produced $7 trillion in sales through their foreign affiliates and are the main force behind worldwide flows of capital, goods, and services. TNCs gain economies of scale through reinforcing their brands and extending their marketing worldwide. Perhaps the best example of a worldwide TNC is Royal Dutch/Shell with more than 70% of its assets and employees based outside the headquarters nation.

Perhaps the best image for the global economy is a network. Integration through networks has resulted in a more resilient world economy with quick mobility of capital flows. A summary of world economy comparisons is given in Table 12.1.

U.S. International export trade has risen from 4.7% of the GDP in 1962 to 11.5% of GDP in 1997. The United States is the world's leader in telecommunications technology and is exploiting it in order to build business networks worldwide. As shown in Table 12.1, the main players in the global economy are the U.S., Europe, and Japan. The challenge of the 21st century will be to foster a more comprehensive spread of economic progress to LDCs [Rosenzweig, 1998]. Communications using the telephone, wireless telephone, and the internet will help to spread global business. The U.S., Canada, and Mexico make up the current NAFTA nations, but that group will extend to other Central and South American nations over the next decade. The potential for the NAFTA Americas is a total population of 455 million.

The best country candidates for prosperous growth are those with stable political and economic policies that foster growth within fiscal and monetary disciplinary policies. The Asian financial crisis that started in the summer of 1997 began, in part, due to currency devaluations in Thailand, Indonesia, and Malaysia. Asset prices and currency prices have since stabilized, and it appears that improvements in these economies are appearing. Japan, China, and the Asian economies are strengthening and may boost the world economy in the next decade.

The emerging economies of Asia include China, Indonesia, Malaysia, and Thailand. The Asian developed economies include Japan, Hong Kong, South Korea, Taiwan, and Singapore.

**Table 12.1   World Economy Comparisons**

|  | Population (Millions) | GDP per Person ($000) | Percent of Total Exports & Imports | Percent of Total World GDP |
|---|---|---|---|---|
| Euro Nations | 289 | 23.8 | 17.5 | 23.2 |
| Other Europe Union | 25 | 219 | 3.0 | 1.9 |
| United Kingdom | 58 | 20.0 | 6.0 | 3.9 |
| United States | 267 | 28.6 | 15.9 | 25.8 |
| Japan | 126 | 36.8 | 8.4 | 15.6 |
| South America | 476 | 3.4 | 6.2 | 5.5 |
| China | 1,200 | 0.7 | 4.5 | 2.8 |
| Africa | 740 | 1.4 | 2.5 | 3.6 |
| India | 940 | 0.4 | 0.8 | 1.1 |
| Other Asia | 1,190 | 1.8 | 23.3 | 7.3 |
| Rest of World | 466 | 5.1 | 11.9 | 9.3 |
| World | 5,777 | 5.1 | 100.0 | 100.0 |

**Source:** *The Babson Staff Letter,* May 29, 1998.

# Investing Globally

The U.S. represents 26% of the world's estimated Gross Domestic Product (GDP) as shown in Table 12.1. The importance of Europe with 29% and Japan with 16 percent of world GDP respectively is clear to all investors. Furthermore, investors are often attracted to emerging countries such as China, Malaysia, and Thailand due to the higher real growth rates for their economies.

Diversification of an investor's portfolio should, ideally, combine solid investments tied to the economies of other regions or nations whose business cycles complement one another. If some zig when others zag, the end result should be higher returns over the long term with lower risk [Malkiel, 1998]. This perfect result doesn't always work out in practice since economies are globally linked. In fact, some economies have a low correlation with the U.S. when the U.S. is doing well. However, these same economies may have high correlations when the U.S. market is falling. Malkiel, in his book *Global Bargain Hunting*, shows that from 1985 through 1995, allocating about 60% of your stocks to the U.S., 25% to other developed markets and 15% to emerging markets provided the highest reward in relation to risk. Unfortunately, if you followed that allocation for 1995–1999, you would have experienced significant disappointments due to poor foreign market performance in Asia and Japan.

For the period 1970–1991, the correlation of the U.S. market with Europe and Japan stocks was measured as about 0.55. However, with increased globalization, that correlation may now be closer to 0.75. The correlation of the United States stock market with the emerging stock markets has been estimated to be 0.5, but the standard deviation (volatility) of returns in the emerging markets is higher (see Table 12.2).

Emerging economy stock markets often are difficult to evaluate. These markets often employ less clear accounting standards, use questionable trading practices, and provide less information to the investor. In general, most investors should avoid direct investment in the stocks of companies in emerging markets. With limited disclosure and risky currencies, most investors will find investments in emerging countries very difficult to analyze and follow. Emerging market returns are quite volatile, often three or four times as volatile as the returns of the U.S. stock market. Table 12.2 shows the average annual return of selected markets for the period 1969–97. The emerging markets of Mexico and Hong Kong provided higher returns, but a volatility three times higher than for the U.S. The Netherlands had a higher correlation with the U.S. market with only a modest increase in volatility.

The average annual return for the United States, Europe, and Japan for the period 1989–98 is shown in Table 12.3. During this period the wisest course of action was to

**Table 12.2   Global Market Returns and Volatility for 1969–97**

| | Average Annual Return (%) | Volatility-Standard Deviation (%) | Correlation with U.S. Market |
|---|---|---|---|
| United States | 14 | 16.8 | 1.0 |
| Hong Kong | 30 | 51.0 | 0.43 |
| Mexico | 25 | 50.0 | 0.18 |
| Netherlands | 19 | 19.0 | 0.75 |

**Table 12.3    The Average Annual Return of Regional Economies 1989–98**

| Region | Annual Return (%) |
|---|---|
| U.S.: S&P 500 | 19.5 |
| Europe | 16.5 |
| Asia without Japan | 7.4 |
| Japan | −4.4 |

**Table 12.4    The Best Asset Allocation of U.S. and International Stocks**

| Period | 1979–88 | | | 1989–97 | | |
|---|---|---|---|---|---|---|
| Allocation | 100% S&P 500 | Best: 50% S&P 50% EAFE | 100% EAFE | 100% S&P 500 | Best: 80% S&P 20% EAFE | 100% EAFE |
| Annual Returns (%) | 16.2 | 19.5 | 22.1 | 18.5 | 16.0 | 4.2 |
| Risk: Std. Deviation (%) | 16.5 | 14.5 | 17.1 | 12.4 | 11.9 | 17.1 |

EAFE: Europe and Far East Index.

invest in the U.S. and Europe. Japanese stocks have been weak for a decade. The return on equity of an average firm on the Tokyo Stock Exchange was only 3.5% in 1998. Furthermore, Japanese firms have debt/equity levels nearly three times those of U.S. firms.

The best asset allocation between U.S. and foreign stocks changes between different periods as shown in Table 12.4. During the period 1979–88, the best allocation was 50% U.S. stocks and 50% foreign stocks resulting in the best return for the risk. On the other hand, for the period 1989–97, the best allocation was 80% U.S. stocks and 20% foreign stocks. Furthermore, the best allocation for 1993–98 was 100% S&P 500 if you were willing to accept the modestly higher volatility.

What about the next decade? Certainly, it will pay to invest a part of our portfolio in foreign stocks or U.S. firms that are very active in foreign markets. Perhaps 10–20% of our portfolio should be invested outside the U.S.

## Selecting Stocks of Foreign Companies

There are excellent companies in Europe, Japan and other industrialized regions that offer good potential to the investor. Some of them are listed on the New York Stock Exchange and provide financial information to many analysts. Other foreign firms are provided to U.S. residents as **American Depository Receipts** (ADRs). An ADR is a negotiable receipt representing the common stock in a foreign corporation. Thus, an ADR is a certificate representing a percentage ownership in the securities of the foreign firm.

Table 12.5 shows several measures of performance for ten foreign companies which are commonly followed by U.S. analysts and investors. Value Line provides a one-page survey on each of these stocks. The measures of performance include average annual returns for the 1993–98 period, projected return on capital and growth rate

**Table 12.5   Ten Foreign Stocks and Associated Measures of Performance Prices as of April 15, 1999**

| Name | Symbol | Stock Exchange | Country | Industry | Average Annual Return 1993–98 (%) | Return on Capital (%) | Growth Rate of Cash Flow (%) | Expected 1999–2003 Power | Value Index (Apr. '99) |
|---|---|---|---|---|---|---|---|---|---|
| ABB | ABBBY | NDQ | Sweden/ Switzerland | Elec. Equip. | 11.3 | 17.5 | 11.5 | 169.1 | 23.8 |
| Aegon | AEG | NY | Netherlands | Insurance | 67.9 | 15.5 | 16.5 | 291.6 | 7.7 |
| Ahold | AHO | NY | Netherlands | Food Retail | 25.9 | 7.5 | 14.0 | 105.0 | 7.5 |
| BP Amoco | BPA | NY | United Kingdom | Petroleum | 24.8 | 13.0 | 8.0 | 124.0 | 9.5 |
| Ericsson | ERICY | NDQ | Sweden | Telecomm. Equipment | 37.6 | 21.0 | 15.5 | 332.0 | 20.1 |
| Nokia | NOKA | NY | Finland | Telecomm. Equip. | 81.3 | 25.5 | 25.5 | 663.3 | 23.7 |
| Pharmacia & Upjohn | PNU | NY | United Kingdom | Drugs | 27.7 | 14.5 | 8.0 | 111.0 | 5.0 |
| Sony | SNE | NY | Japan | Movies and Electronics | 5.0 | 7.0 | 13.0 | 85.0 | 11.5 |
| Toyota | TOYOY | NDQ | Japan | Autos | 7.1 | 8.0 | 9.0 | 75.0 | 7.1 |
| Vodaphone | VOD | NY | United Kingdom | Telecomm. Services | 46.4 | 24.0 | 24.5 | 599.8 | 21.4 |

of cash flow. Using these and other figures provided by Value Line we use the equations developed in Chapters 7 and 8 for Power and the Value Index to calculate these measures for each stock.

Asea Brown Boveri (ABB) is a European-based firm justifiably respected for its global accomplishments. ABB is engaged in electrical equipment and electric power generation equipment. ABB's Power and Value Index are reasonably attractive.

Aegon (AEG) is a transnational insurance company. AEG purchased the insurance operations of the U.S.-based Providian Company in 1996. Aegon is primarily active in Europe and the U.S. The average annual return was 68% for the period 1993–98. Aegon announced in February 1999 that it was purchasing Transamerica Corp., a U.S. life insurance company. We use return on equity and growth of earnings to calculate Power for this financial stock. The Power of the stock is solid, but the Value Index is low due to a high PE ratio equal to 40 at the time of this writing. This may be an attractive stock if the PE declines to 30 or the return on equity increases significantly.

Ahold is an international food retailer with more than 3,500 grocery stores throughout the world. It has a low power and value index and would be a weak candidate for purchase.

BP Amoco is a large and growing oil and gas exploration and production company. It has a modest Power and Value Index. This firm is a world leader in its industry and is a potential candidate for purchase in order to obtain long-run returns in the energy industry.

Pharmacia & Upjohn is a pharmaceutical company formed from the 1995 merger of the Swedish Pharmacia and the U.S.-based Upjohn Co. The company has gone through a turnaround in 1998–99 and may offer moderate returns in the next five years, if growth in cash flow can increased.

Sony is a large Japanese firm which produces movies and provides consumer electronic products. Like many Japanese companies its return on equity and return on capital is low. Sony can only provide modest returns in the future.

Toyota is a Japanese auto firm that offers the popular Camry and Lexus. It is an excellent company but offers modest returns to shareholders.

The best opportunities in foreign stocks are in the telecommunications industry which is benefiting from worldwide growth. Ericsson, Nokia, and Vodaphone are telecommunications equipment and services providers. All three firms have good measures of Power and Value Index. Ericsson is active in more than 100 countries. Nokia has solid earnings growth and an above average Value Index. Vodaphone is the largest mobile telecommunications provider in Britain and has recently merged with the U.S. mobile phone provider AirTouch Communications.

If we pick one or two large foreign stocks, the choice would be Nokia and Vodaphone.

## Selecting U.S.-Based Transnational Companies

The easiest approach to stock investing in foreign markets is to select U.S.-based companies with a large percent of their sales in the international markets. All necessary information about U.S. companies is readily available and the cost of purchase is modest. Furthermore, the investor has limited exposure to currency risk. We will concentrate on global leaders whose products and services are well-known to Americans and are becoming increasingly familiar to foreigners and that promise long-term growth and aren't overvalued.

Table 12.6 lists 23 U.S.-based transnational companies with significant foreign sales. An investor can readily build a portfolio from this list and obtain significant exposure to foreign markets. Coca-Cola obtains 66% of its sales outside the U.S. and is one of the world's leading brand names. Growth of cash flow should increase as the Asian financial crisis moderates.

Disney is the world's leading media and entertainment company. Its unique combination of theme parks, consumer products, motion pictures, and broadcasting provides great power worldwide.

Procter and Gamble sells more than 300 brands of food, health care, and cleaning products in 150 countries. They are striving to plan, develop and manage all lines of products on a global basis.

Examples of great worldwide companies that have stumbled but may regain momentum are Boeing, Motorola, and Minnesota Mining. These firms many turn their fortunes in the near future.

The most attractive U.S. transnationals include Avon, General Electric, IBM, Intel, and Microsoft. All these firms have a high Power and Value Index and have provided significant returns in the past. A portfolio that included these firms would benefit from significant exposure to foreign markets.

**Table 12.6    Twenty-three U.S.-Based Transnational Companies with Attractive Performance**

| Name | Sales: Percent Foreign Total | Industry | Average Annual Returns 1993–98 (%) | Expected 1999–2003 Return on Cap. (%) | Expected 1999–2003 Growth Rate of Cash Flow | Value Power | Value Index | Value Line Safety Rating | Operating Margin (%) |
|---|---|---|---|---|---|---|---|---|---|
| Aflac | 80 | Insurance | 37.3 | 13.0 | 13.0 | 185.9 | 21.4 | 3 | NA |
| Amer Int'l Gp | 50 | Insurance | 30.5 | 13.5 | 15.5 | 272.0 | 16.1 | 2 | NA |
| Avon | 66 | Cosmetics | 32.8 | 72.5 | 15.0 | 1136.4 | 49.6 | 3 | 17.0 |
| Boeing | 47 | Aircraft | 10.2 | 20.0 | 17.0 | 204.0 | 21.7 | 2 | 12.0 |
| Coca-Cola | 66 | Beverage | 26.1 | 45.0 | 9.0 | 510.3 | 14.2 | 1 | 31.0 |
| Colgate | 67 | Home Products | 27.1 | 24.0 | 12.5 | 378.0 | 18.6 | 1 | 22.0 |
| Daimler Chrysler | 45 | Autos | NA | 9.0 | 6.0 | 54.0 | 9.0 | 3 | 13.0 |
| DuPont | 49 | Chemicals | 20.2 | 19.0 | 4.5 | 111.0 | 10.1 | 1 | 26.0 |
| Exxon | NA | Petroleum | 22.6 | 15.0 | 5.0 | 73.5 | 5.25 | 1 | 14.0 |
| Ford | 34 | Autos | 27.8 | 8.0 | 7.0 | 56.0 | 12.4 | 3 | 25.0 |
| General Elec. | 48 | Elec. Equip. | 34.2 | 26.0 | 13.5 | 491.1 | 19.6 | 1 | 22.0 |
| Gillette | 62 | Home Products | 27.7 | 23.5 | 13.5 | 391.8 | 13.3 | 2 | 29.0 |
| Hewlett-Packard | 56 | Computer Equip | 29.5 | 17.0 | 12.5 | 282.6 | 20.5 | 3 | 12.5 |
| IBM | 46 | Computer Equip | 47.1 | 30.0 | 12.5 | 472.5 | 37.2 | 2 | 17.5 |
| Intel | 56 | Semiconductors | 50.6 | 26.0 | 14.0 | 484.1 | 24.2 | 3 | 48.0 |
| Johnson & Johnson | 48 | Medical | 32.4 | 26.5 | 12.0 | 378.4 | 13.2 | 1 | 27.5 |
| McDonalds | 50 | Restaurants | 22.7 | 13.0 | 10.0 | 160.6 | 8.0 | 1 | 32.0 |
| Microsoft | 38 | Software | 68.9 | 28.0 | 27.0 | 1005.5 | 19.8 | 2 | 60.0 |
| Minnesota Mining | 52 | Materials | 9.4 | 22.0 | 8.0 | 184.8 | 15.7 | 1 | 24.0 |
| Motorola | 58 | Communicat. Technology | 6.6 | 12.0 | 9.0 | 90.7 | 7.6 | 3 | 16.0 |
| Pfizer | 45 | Drugs | 51.2 | 34.0 | 17.5 | 708.1 | 17.7 | 2 | 34.0 |
| Procter & Gamble | 35 | Home Products | 28.6 | 23.7 | 13.0 | 431.3 | 18.7 | 1 | 23.0 |
| Xerox | 49 | Copiers and Printers | 34.6 | 16.0 | 12.5 | 216.0 | 16.6 | 3 | 21.0 |

**Prices for Period:** April 1–18, 1999.

# 13 Technology Stocks

## This chapter covers

- Innovation and Technology
- Valuation of Technology Companies
- Technology Stock Selection
- Emerging Internet Stocks
- Biotechnology

## Innovation and Technology

**Innovation** is the creation and introduction of a new technology, tool or process which is a better way of achieving a task or objective. A list of innovations is provided in Table 13.1. An entrepreneur or entrepreneurial firm introduces an innovation and receives the return on the investment. Innovations not only break new ground, they also yield far better returns than ordinary business ventures. Our world economy is not in equilibrium, but rather it is being constantly disrupted by technological innovation. These innovations come in waves of change as one innovation begets another. Table 13.2 lists a model of the five waves of innovation. The current

**Table 13.1 Innovations in History**

| Period of Introduction | Innovation |
|---|---|
| 1280–1330 | Eyeglasses |
| 1420–1440 | Navigation Technology |
| 1450–1460 | Gutenberg's Moveable Type |
| 1700–1780 | The Factory System |
| 1800–1840 | Railroads |
| 1880–1900 | Internal Combustion Engine |
| 1880–1900 | Water Purification Systems |
| 1880–1910 | Electric Power Systems |
| 1920–1940 | Antibiotics |
| 1945–1960 | Digital Computer |
| 1950–1960 | Transistor |
| 1960–1970 | Integrated Circuit |
| 1998– | Digital Television |

**Table 13.2   A Model of the Five Waves of Technological Innovation**

| Wave | Time Period | Technologies |
|------|-------------|--------------|
| 1 | 1785–1845 | Water Power, Textiles, Iron |
| 2 | 1845–1900 | Steam Power, Railroads, Steel |
| 3 | 1900–1950 | Electric Power, Chemicals, Internal Combustion Engine |
| 4 | 1950–1990 | Electronics, Aviation, Petrochemicals |
| 5 | 1990–2030 | Digital Networks, Software, and New Media |

wave of digital networks and media is providing the investor with great opportunities to reap the benefits of change. The current wave is based on semiconductors, fiber optics, software, computers, and information technology.

Federal Reserve Chairman Alan Greenspan said on May 6, 1999, that an unexpected leap in technology is primarily responsible for the nation's "phenomenal" economic performance and the extraordinary combination of strong growth, low unemployment, low inflation, high corporate profits and soaring stock prices.

It is clear that intellect, innovation, and technology are central to economic growth. Intellectually based services and products provide 70% of all new jobs and 75% of all the value added in the economy. About 30% of U.S. economic growth comes from technology and that may increase over the next decade with the ever-expanding internet and the growth of telecommunications throughout the world. The internet is an innovation that has reached 25% of the households and 50% of the businesses in seven years (1992–99).

Some high-tech industries have limited pricing capabilities. Consider the pricing pressure on computers. Although the unit sales of personal computers are expected to increase 14% in 1999, the revenues may only increase 5% or less. Furthermore, many companies are finding they are falling behind in the e-commerce (internet) business channel for their industry and rushing to catch up.

The telecommunications equipment industry is growing fast and retains good pricing power. Telephone and computer networks are spreading worldwide and fueling demand for telephone circuit switches that route billions of phone calls each day. In addition, the growth of data communication equipment is growing at a high rate. Data communications use packets of data (segments) allowing communication lines to carry multiple messages. The global telecommunications network is the largest machine (network) ever built. The dominant companies include Lucent, Cisco, and Tellabs.

## Valuation of Technology Companies

The best business has an impregnable franchise with clear and sustainable advantages. Of course, all these business are subject to technological change and may miss the next wave of change. Exhibit 13.1 lists the traits of a successful technology company. The greatest risk for a technology company is the diminishment of the value of the current product series by a new technology or product substitution.

More than in any sector, technology stocks tend to trade on earnings momentum, and less on fundamental valuation methods. Technology, by its nature, is forward

**Exhibit 13.1    Traits of a Successful Technology Company**

- Significant barriers to entry
- High customer switching costs
- Proprietary products
- Focused, aggressive management
- Low pricing pressure
- Attractive industry structure

looking and estimates of significant growth are normal. In a way, the investors in the market sense the growth of new technology sectors and bid up the growing companies. The question of sustainability of strong growth remains unensured. Technology companies have returned an average annual return of 21.9% over the 10%-year period 1989–98. However, no sector reigns forever—technology stocks returned a negative 1.5% over the four-year period 1984–88. Technology stocks are relatively uncorrelated with other sectors as shown in Table 13.3. Since technology stocks are purchased for their growth prospects, they are relatively uncorrelated with value stocks, utilities, and real estate. Since we hold four subportfolios, we have significant diversification and do not count on only technology growth and returns. We include high growth technology companies, such as Microsoft, in our Gazelle portfolio. Lower growth, steady performers, like IBM and General Electric, are included in our Hare portfolio.

Table 13.4 shows the subsectors of the technology sector that we are interested in. We show the estimated market size and selected leading companies in each sector. All these markets are growing at 15% or better.

**Moore's Law,** created by Gordon Moore, cofounder of Intel, states that the performance of integrated circuits (ICs) will double every 18 months. Furthermore, an estimate of the growth of the speed of communication systems, bandwidth, is doubling every 12 months. These performance changes are leading to expansion of data networks, video conferencing and the internet.

The technology sector is, by its nature of vast change, very volatile and risky. For example, the average annual return of a technology sector mutual fund is about 35% more volatile than the S&P 500. We can moderate that volatility by holding a broad, diversified portfolio of which technology may be only 10–20% of the total portfolio.

Technology companies and their stock prices are very vulnerable to industry shifts. Cadence Design Systems, the leader in semiconductor circuit design software tools, experienced difficulty in transitioning to the 0.18-micron technology and experienced flat sales and profits. With that announcement, anxiety in the marketplace cut their share price by 40% in two days.

When valuing a technology stock, one must recognize that a significant part of a technology company's assets are intellectual capital which has been generated through R&D investments. General accounting principles record R&D as an expenditure, but

**Table 13.3    Correlation of Technology Stocks with Other Sectors**

| Sector | Real Estate | Utilities | Natural Resources | Large Value | Large Growth |
|---|---|---|---|---|---|
| Correlation with Technology | –13% | 26% | 29% | 55% | 81% |

**Table 13.4  Sub-Sectors of the Technology Sector**

| Subsector | Description | Annual Market Size ($ billion) | Growth Rate of Sales (%) | Leading Firms |
|---|---|---|---|---|
| Personal Computers and Equipment | Desktop and laptop, storage, printers | 240 | 16 | IBM, Dell, Compaq, HP, EMC |
| Computer Networks and Internet | Servers and network software | 100 | 30 | Cisco, 3Com, Sun |
| Software | Programs and codes | 140 | 20 | Microsoft, Oracle, Computer Associates |
| Semiconductors | Integrated circuits | 150 | 18 | Intel, Texas Instruments, Xilinx |
| Semiconductors— Capital Equipment | Process equipment for manufacturers | 25 | 15 | Applied Materials, Terradyne, Novellus |
| Telecommunication Equipment | Telecomm. system equipment | 200 | 15 | Lucent, Nokia, Motorola, Tellabs |
| Computer-Aided Design | Software + hardware for technology design | 50 | 15 | Parametric Tech., Synopsys, Autodesk, Cadence |
| Medical | Medical equipment | 100 | 15 | Medtronic, Guidant |

we recognize it as an investment. Most technology companies spend 7% to 12% of sales revenues on R&D. The R&D expenditures as a percent of sales for selected large companies is shown in Table 13.5. In general, we are looking for firms with R&D exceeding 7 percent of sales. In addition, we are seeking sales growth greater than 15% and a return on sales (profit margin) of 10% or greater. We will calculate the Power

**Table 13.5  Research and Development Expenditures of Large Technology Companies—Fiscal Year 1977**

| Name | R&D ($ millions) | R&D (% of sales) |
|---|---|---|
| IBM | 4,877 | 6.2 |
| Hewlett-Packard | 3,078 | 7.2 |
| Lucent | 3,023 | 11.5 |
| Motorola | 2,748 | 9.2 |
| Intel | 2,347 | 9.4 |
| Microsoft | 1,925 | 16.9 |
| Texas Instruments | 1,536 | 15.8 |
| Xerox | 1,079 | 5.9 |
| Sun Microsystems | 849 | 9.9 |
| Compaq Computers | 817 | 3.3 |
| Cisco | 698 | 10.8 |
| Applied Materials | 568 | 13.9 |
| Oracle | 556 | 9.8 |

**Exhibit 13.2    Valuation Criteria for Technology Stocks**

- R&D exceeds 7% of sales
- Predicted growth rate of sales exceeds 10%
- Predicted return on sales (profit margin) exceeds 10%
- Power exceeds 400
- Value Index exceeds 16
- Value Line safety rating of 3 or better

and the Value Index and purchase companies with a Value Index exceeding 20. These valuation criteria are summarized in Exhibit 13.2.

## Technology Stock Selection

We are seeking outstanding technology stocks for our portfolio using the criteria of Exhibit 13.2. Table 13.6 provides a list of 21 technology stocks that satisfy at least five of the six criteria of Exhibit 13.2. The Value Line chart for Microsoft is shown in Figure 13.1. Note the outstanding performance of this company with its excellent consistent growth rate.

Based on the Value Index, the best companies described in Table 13.6 are BMC Software, Dell, IBM, and Oracle since all have a Value Index greater than 36. Parametric Technology is also attractive due to a relatively low PCF. Microsoft has the highest PCF with a value of 53.3 (excluding AOL with PCF = 236).

AOL has a PCF of 236 and a high power of 2,244. AOL is a special case that we consider along with other internet stocks in the next section.

Note that the PCF/G ratio for these stocks ranges from a low of 0.8 to a high of 2.2. The price-to-sales ratio ranges from a low of 1.4 for HP to a high of 22.9 for Microsoft. The average price-to-sales ratio for the 20 (excluding AOL) stocks is 7.0. Many of the stocks in Table 13.6 appear to be overvalued since their actual PCF is from 50% to 100% higher than their normal Value Line PCF. Almost every year a technology stock will run ahead of their normal price trajectory, but at other times of the year they may more closely approach their Value Line normal PCF. The time to buy is when the PCF approaches the normal Value Line PCF.

## Emerging Internet Stocks

Fast-growing intranets and the internet now connect a variety of companies and individuals in a web of electronic commerce. The internet is a global collection of computer networks permitting individuals and institutions to share information. It is estimated that 140 million people are connected on-line and that internet traffic is doubling every four months. The percentage of U.S. households that regularly log on to the internet each week now approaches 25%. Businesses increasingly use e-mail, file transfer and other data applications. On-line shopping in the U.S. may grow from $8 billion (estimated) in 1999 to $40 billion in 2003–04. On-line trading of stocks now accounts for about 15% of all trades and is growing at 30% per year. The impact of the internet on the nation's economy is often understated. It was responsible for 1.2 million jobs and $300 billion in revenue in 1998, according to a study by the University of

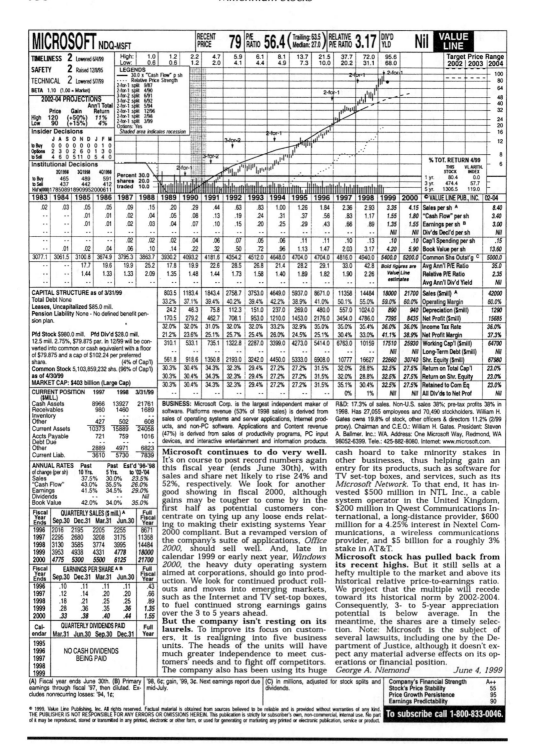

Figure 13.1   Courtesy of Value Line Publishing Inc.

**Table 13.6  Selected Measures for 21 Technology Stocks**

| Name | Sales 1999 ($ Mill) | 1999–2003 Growth Rate of Sales (%) | Profit Margin 1999 (%) | 1999–2003 Return on Capital (%) | 1999–2003 G = Growth Rate of Cash Flow (%) | Value Line Normal PCF | PCF | R&D Sales (%) | Power | Value Index | PCF G | Price Sales (%) | 1993–1998 Average Annual Return (%) |
|---|---|---|---|---|---|---|---|---|---|---|---|---|---|
| America OnLine | 4,320 | 36.0 | 9.0 | 33.0 | 68.0 | NA | 236 | NA | 2,244 | 9.5 | 3.5 | 28.0 | 143.1 |
| Applied Material | 3,200 | 14.5 | 10.0 | 18.0 | 14.5 | 13.5 | 31.5 | 16.0 | 305 | 9.7 | 2.2 | 5.5 | 34.5 |
| BMC Software | 1,325 | 25.5 | 33.2 | 25.0 | 25.0 | 20.0 | 13.2 | 13.0 | 584 | 44.2 | 0.5 | 5.1 | 49.3 |
| Compaq | 42,000 | 22.0 | 6.9 | 19.0 | 20.0 | 12.5 | 15.0 | 3.3 | 426 | 28.4 | 0.8 | 1.2 | 53.6 |
| Cisco | 11,300 | 28.0 | 21.7 | 27.0 | 28.0 | 25.0 | 61.0 | 12.0 | 1,058 | 17.3 | 2.2 | 15.6 | 66.8 |
| Dell | 25,000 | 47.0 | 7.8 | 60.0 | 50.0 | 22.0 | 47.1 | 1.7 | 3,990 | 84.7 | 0.9 | 3.8 | 152.9 |
| EMC | 5,225 | 26.0 | 20.7 | 23.0 | 30.0 | 30.0 | 47.0 | 7.5 | 852 | 18.1 | 1.6 | 11.9 | 59.4 |
| HP | 51,000 | 13.5 | 6.8 | 17.0 | 12.5 | 12.0 | 13.5 | 7.2 | 283 | 20.9 | 1.1 | 1.4 | 29.5 |
| IBM | 89,000 | 14.0 | 7.6 | 29.5 | 12.5 | 7.0 | 12.3 | 6.2 | 465 | 37.8 | 1.0 | 1.7 | 47.1 |
| Intel | | 13.5 | 26.3 | 28.0 | 14.0 | 13.0 | 19.0 | 9.4 | 521 | 27.4 | 1.4 | 7.0 | 50.6 |
| Lucent | 34,900 | 12.5 | 9.0 | 24.0 | 19.0 | 26.0 | 34.3 | 12.2 | 593 | 17.3 | 1.8 | 4.6 | 87.8* |
| Linear Tech | 500 | 18.5 | 38.5 | 21.0 | 19.5 | 20.0 | 41.4 | 9.5 | 467 | 11.3 | 2.1 | 17.6 | 36.5 |
| Medtronic | 5,625 | 11.5 | 23.3 | 12.0 | 13.0 | 18.0 | 25.5 | 11.4 | 183 | 7.2 | 2.0 | 8.6 | 49.5 |
| Microsoft | 18,300 | 23.5 | 37.3 | 28.0 | 27.0 | 30.0 | 53.3 | 17.3 | 1,005 | 18.9 | 2.0 | 22.9 | 68.9 |
| Nokia | 20,500 | 26.0 | 11.6 | 29.0 | 25.5 | 14.0 | 28.1 | 9.0 | 754 | 26.8 | 1.1 | 4.3 | 76.1 |
| Oracle | 9,100 | 26.5 | 13.7 | 33.0 | 22.5 | 25.0 | 21.8 | 10.1 | 820 | 37.6 | 1.0 | 3.8 | 38.3 |
| Parametric Tech | 1,065 | 22.5 | 22.0 | 34.0 | 19.5 | 22.0 | 17.9 | 9.2 | 636 | 35.6 | 0.9 | 4.3 | 10.9 |
| Sun | 11,350 | 16.0 | 9.2 | 22.5 | 16.0 | 13.0 | 14.8 | 10.4 | 398 | 26.9 | 2.0 | 2.3 | 63.7 |
| Texas Instr. | 9,000 | 6.0 | 12.2 | 15.0 | 13.5 | 9.5 | 18.2 | 11.0 | 237 | 13.0 | 1.3 | 4.3 | 0.2 |
| Tellabs | 2,150 | 21.0 | 22.3 | 23.0 | 24.0 | 20.0 | 36.4 | 12.2 | 662 | 18.2 | 1.5 | 9.2 | 63.3 |
| Xilinx | 770 | 10.0 | 21.8 | 16.0 | 12.0 | 15.5 | 15.7 | 13.0 | 158 | 10.1 | 1.3 | 4.3 | 32.5 |

*Two years
Average Prices for Period April 1–22, 1999.

Texas. The internet is fast becoming part of the mainstream economy. You can follow the growth of the internet at www.internetindicators.com. The internet permits customized messages and global reach. It is a high fixed-cost, low marginal-cost business. This creates scale economics and may lead to "increasing returns" if one provider becomes the standard. (See Chapter 2 for a discussion of increasing returns.)

Networks, such as the internet, follow **Metcalfe's Law** which states that the value of a network, V, increases with the square of the number of users, n. Therefore,

$$V = kn^2$$

where k = constant. As the number of network users increases and the cost per user decreases, the network becomes very powerful. Thus, we expect the internet will transform markets, unseat market leaders and provide large returns to early investors.

The issue of barriers to entry becomes crucial when you consider that the valuations of internet stocks assume that the current leaders will retain their competitive advantages for years to come. After all, with no profits to speak of in the present, virtually all the value of most internet stocks lies in the future, and that future value depends quite a bit on what sort of growth assumptions you use.

Emerging companies that are leaders in electronic commerce can become the successful leaders of tomorrow. It is difficult to pick the winners in the early stages since they usually have little, if any, earnings. AOL is a potential winner, providing an average annual return of 143 percent over the five-year period 1993–98. In 1991, AOL had $21 million in sales and 181,000 subscribers. By 1999 it had sales of $3 billion and 18 million subscribers. AOL survived the competition with CompuServe, Prodigy, and MSN through good management, aggressive marketing, and competitors' mistakes.

We will consider the six stocks based on the Internet as listed in Table 13.7. Since most of the firms do not have significant earnings we use the growth rate of revenues (sales) as the basic power factor. Thus, we define Power for Internet stocks as

$$\text{Power} = G \times \text{Rcap}$$

**Table 13.7    Performance Measures for Six Stocks Based on the Internet**

| | Industry | G = Growth Rate of Revenues (%) | Return on Capital (%) | 1999 PCF | Power | Value Index | 1999 Price Sales | Estimated 1999 Sales ($ Mill) |
|---|---|---|---|---|---|---|---|---|
| | | 1999–2003 | | | | | | |
| Amazon.Com | Books and music | 51.0 | 11.5 | NM | 587 | NM | 23 | 1,200 |
| America OnLine | Information services | 36.0 | 37.5 | 276 | 1,350 | 4.9 | 33 | 4,230 |
| Cnet | Web content | 60.0 | 18.0 | 192 | 1,080 | 5.6 | 40 | 100 |
| E-Trade | Stock broker | 60.0 | 22.5 | 500 | 1,350 | 2.7 | 23 | 500 |
| Schwab | Stock broker | 17.5 | 18.0 | 120 | 315 | 2.6 | 14 | 3,700 |
| Yahoo | Web search | 70.5 | 21.5 | 354 | 1,516 | 4.3 | 111 | 375 |

Prices of April 15–23, 1999.

where G = growth rate of revenues and Rcap = return on capital for the period 2001–2003 as projected by Value Line. The prices of internet stocks normally track the revenue growth of the companies.

For AOL we have

$$\text{Power} = 36.0 \times 37.5 = 1{,}350$$

The price to cash flow for 1999, using an average price for April 15–23, 1999, is a very large 276. Thus, the Value Index is a very modest 4.9.

In a similar way we calculate the power for Amazon.com. Amazon has negative cash flow, so PCF is undefined. Amazon is the leading on-line direct seller of books plus CDs, videotapes, and audiotapes. Since opening its virtual doors in July 1995, the company has attracted a customer base of 5 million accounts.

Yahoo is a information content provider and search engine. Yahoo has become known as a gateway (portal) and has about 40 million registered users. Yahoo appears to have good growth and staying power. Its price to cash flow, at 354, is not useful as a measure of value. Similarly, the price to cash flow of the online securities broker, E-Trade, at 500, is of little meaning. Nevertheless, E-Trade has generated 9 million accounts by April 1999 and is growing revenues at 60% per year.

Schwab, the securities broker, was generating 50% of its transactions online in 1999 and thus obtained much of the high valuation (and potential) of internet stocks.

The valuation of internet stocks may be calculated using the growth of revenues which may be considered a proxy for the intangible future value of these companies. These companies are racing to get market share and build up a subscriber list of satisfied customers. Purchasing a book from Amazon or a stock through Schwab is a productive and satisfying method for many.

Valuing companies with erratic earnings—or no earnings at all—can be difficult. That's where the price-to-sales ratio comes in. Unlike the more common price-to-earnings ratio, price to sales can be used to value any public company, and it sometimes provides a more stable valuation measure.

A company's price-to-sales (PS) ratio is calculated by taking its current share price and dividing by its estimated revenues per share for the year under consideration.

One method of valuation is to compare the price-to-sales ratio of very successful technology companies. Examining Table 13.6, we find Microsoft has a PS ratio of 23 and Cisco has a PS equal to 16. The PS ratios for Amazon.com, Schwab, and AOL are similar in magnitude. Thus, investors' expectations are high for these new companies to become the next Microsoft.

Let's develop a PS model based on the expected growth rate of revenues, g, and a discount rate, r. Then, the price is calculated as

$$\text{Price} = 1999 \text{ Sales} \times \left(\frac{1+g}{1+r}\right)^{N} \times \text{PS}(N)$$

where we use N years and PS(N) is the price-to-sales ratio in the Nth year. We will use r = 0.16, g = 0.40, and N = 4 years. This assumes a growth rate of 40% for four years which very few firms have attained. Also, we use Cisco's price-to-sales ratio, $PS(N) = 16$. Then, we have

$$\text{Current PS} = \left(\frac{1.40}{1.16}\right)^6 \times 16 = 2.1 \times 16 = 33.6.$$

Thus, if you accept a growth rate of 40% for AOL, Amazon, and E-Trade, the value of these stocks may not be excessive. On the other hand, the PS ratio of Yahoo seems unreasonable unless you believe that Yahoo can achieve a growth of revenues equal to 70% for the next four years. Using our formula for Yahoo with $g = 0.70$, we have

$$PS = \left(\frac{1.70}{1.16}\right)^4 \times 16 = 4.6 \times 16 = 74.$$

The Yahoo $PS = 111$, seems excessive. Furthermore, if you consider the riskiness of actually achieving 70% growth rate is higher, you would probably use a larger discount rate, r, to compensate for the higher risk.

When an investor purchases a stock with a PCF or PE greater than 100 or a Price to Sales ratio greater than 15, the investor may be considered to be purchasing an option on the future success of this company. The investor expects very great rewards while accepting very high risk of loss of the investment. The idea is to take an option in several potential winners and thus spread the risk. When the investor identifies a sector such as the internet or biotechnology with great potential it is probably prudent to use a selection strategy similar to one used by venture capitalists. When the future is highly uncertain, it pays to have a broad range of options open. Traditional methods undervalue a company such as Amazon.com that's losing money now because it's spending lots of money creating options for the future.

We propose the High Potential Strategy as outlined in Exhibit 13.3. This strategy accepts the high risks of the emerging companies and attempts to mitigate this risk by diversifying this small subportfolio. The benefits of a large gain from the winner can be significant. As an example, consider a portfolio of three internet stocks purchased in 1997 as shown in Table 13.8. The return to the purchaser is truly amazing. You can use the same strategy in the future, if you determine that the internet or biotechnology sector will emerge as a future source of high growth. Another example of a potential high growth emerging sector is the data networking technology sector. Examples of companies in the data networking sector are Broadcom, PMC-Sierra,

### Exhibit 13.3   The High Potential Strategy

- Identify a new sector that is emerging into high revenue growth greater than 40% annually.
- Select three or four companies that have a clear potential to be the leader in this new area.
- Purchase the stocks of these companies when the price declines to a price to sales ratio less than 25.
- As the clear leader emerges, sell the laggards and consolidate the funds in the leader.
- Hold the leader for the long term.

**Table 13.8   A Portfolio of Four Internet Stocks Purchased in 1997**

| Name | Average* Price ($) April 1997 | Average Price ($) April 1999 | Gain (%) |
|---|---|---|---|
| Amazon.com | 4.0 | 170 | 4150 |
| America OnLine | 6.0 | 150 | 2400 |
| Netscape | 24.0 | 75 | 212 |
| Yahoo | 4.0 | 195 | 4775 |

*Split adjusted;   Price for Amazon.com for May 1997.

Vitesse Semiconductor, and Uniphase. All of these stocks have a PE ratio greater than 50 and an expected growth rate of sales greater than 40%.

# Biotechnology

New research techniques are accelerating the pace of discovery in biology, driving forward medicine and agriculture. Biotechnology focuses on the pharmaceutical and agricultural industries. The biotechnology sector is highly fragmented with more than 1,000 companies. Due to heavy R&D spending in the 1980s and early 1990s the trend is now moving towards applications and drugs for the trial phase. Bolstered by a friendlier regulatory and legislative environment that should shorten the time needed to commercialize products, industry revenues should double over the next five years.

Sales in the biotech industry are expected to grow at a 20% annual rate. However, many investments in biotech firms have not paid off. Biotech companies are difficult for individuals to analyze and understand. It takes about 10 years and $300 million to develop one drug and only one in five will ultimately receive FDA approval.

The performance measures for eight biotech stocks are listed in Table 13.9. We calculate the Power using the formula

$$\text{Power} = G \times Rcap$$

where G = growth rate of cash flow and RCap = estimated return on capital for 2001–03. If Value Line is unable to provide the projected growth of cash flow, we use the growth rate of revenues. Most of these firms have large PCF ratios. We also provide the Price to Sales ratio as an indicator of value.

Monsanto is an agriculture biotech firm with an established product line. The estimated growth of cash flow is low, but the firm may accelerate its products development.

DuPont, the chemical company, has purchased Pioneer Hi-Bred, a large biotech seed company. DuPont will be shifting to biotech with a potential for one-third of its revenues originating from biotech.

The firms with the best record and potential are Agouron, Amgen, and Biogen. It may be prudent to purchase all three and hold them as a biotech portfolio with expected solid long-term results. This sector is risky and any purchaser must be prepared to accept volatile swings as investors' mood changes about this sector occur.

**Table 13.9**   **Performance Measures for Eight Biotechnology Stocks**

| Name | 1999 Sales ($ Mill) | G = Growth Rate of Cash Flow (%) | 2001–03 Return on Capital (%) | Power | PCF | Value Index | Price Sales | Debt Capital | 1993–98 Average Annual Return (%) |
|---|---|---|---|---|---|---|---|---|---|
| Agouron | 610 | 28.5* | 21.5 | 613 | 30.5 | 20.1 | 2.9 | 2 | 58.5 |
| Amgen | 2,895 | 14.5 | 19.5 | 283 | 27.9 | 10.1 | 10.3 | 9 | 33.4 |
| Biogen | 735 | 36.0 | 16.0 | 576 | 31.1 | 18.5 | 9.5 | 8 | 33.0 |
| Genentech | 1,185 | 16.5 | 10.5 | 173 | 34.9 | 5.0 | 8.8 | 6 | 9.6 |
| Genzyme | 825 | 16.5 | 13.0 | 215 | 19.1 | 11.2 | 4.5 | 25 | 30.6 |
| Human Genome | 32 | 39.0* | 13.5 | 527 | NM | NM | 25.7 | 1 | 18.5 |
| Immunex | 425 | 31.5* | 20.5 | 646 | 113 | 5.7 | 14.0 | 3 | 50.6 |
| Monsanto | 9,000 | 8.5 | 12.5 | 106 | 24.2 | 4.4 | 3.0 | 56 | 31.2 |

*Growth rate of Sales.
Average Prices of March 15–April 15, 1999.

# 14 Value Stocks

## This chapter covers

- Rational Investors
- A Value Method
- Selected Value Stocks
- Setting a Strike Price for a Value Stock
- Socially Conscious Investing

## Rational Investors

The **efficient market theory** holds that the price of a stock accurately reflects all currently available information about its fundamental value and therefore it is impossible for the average investor to get an edge over the market. The price of a stock is a function of three factors and may be represented by the equation

$$Price = w_1 r + w_2 g + w_3 e$$

where r is the discount rate, g is the expected growth rate of cash flow (or earnings), e is the emotional factor, and $w_i$ are the weights for each factor. If we neglect the emotional factor, we have the standard discount model (see Chapter 6).

$$Price = CF(1) \sum_{n=1}^{N} x^n$$

where $x = (1 + g)/(1 + r)$ and CF is cash flow.

What rational theory of human behavior can explain wide mood swings in the market resulting in a PCF of 100 for Yahoo and a PCF of 6 for Ford? Many have a remarkable wave of hope for the internet and complete disillusionment for automobile manufacturers. Investors are not always rational and neither are markets, which are communities of investors.

The marketplace of companies doing business globally is very dynamic and competitive. By its nature, a dynamical system is inherently unpredictable. Thus, our estimates of growth rates and appropriate discount rates are subject to continual reassessment. In the short run psychology drives the stock market, while in the long run earnings drive it.

Market participants are subject to the emotions of fear and greed. Investors are scared of losing money and are also anxious to make more profits. Of the two emotions, fear is by far the stronger, as evidenced by stock prices which fall faster than they go up. When fear takes over our emotions magnify the situation.

Investors tend to be overconfident about their forecasts and thus tend to trade excessively. This short-term activity is wasteful and will normally result in below normal returns if one accounts for all the transaction costs (taxes, fees, and price spreads).

In this chapter we seek long-term results by selecting long-term companies in unglamorous industries such as foods, soaps, autos, and beverages. We use a value-based selection method that we expect to provide average returns of 12–14% over a five- to 10-year period. We will develop a method of finding value that will provide good long-run returns by selecting stocks in unglamorous, low-growth industries. We try to distinguish companies that are relatively inexpensive.

## A Value Method

We seek stocks with excellent value which we define as reasonable growth of cash flow and solid, consistent long-term returns. Again, we use the power formula as

$$\text{Power} = \text{Rcap} \times \text{G} \times \text{Sus}$$

where Rcap = return on capital and G = growth rate of cash flow. Here we develop a revised definition of sustainability (Sus) that is reflective of moderate growth stocks as follows

$$\text{Sus.} = \text{Fin. Str. Rating} \times \left( \frac{\text{Price Stability or Price Growth Persistence}}{80} \right)$$

where we use Financial Strength Rating (See Table 7.6) and the larger of the Price Growth Persistence or the Price Stability Index (See Exhibit 7.1). Then, as before, we define the Value Index as

$$\text{Value Index} = \frac{\text{Power}}{\text{PCF}}$$

where PCF = price to cash flow estimated for the coming or current year.

Our selection criteria is shown in Exhibit 14.1.

In this chapter we seek value at a reasonable price. Thus, the price-to-cash-flow ratio is normally less than 15 and often less than 10. Furthermore, since these are less

### Exhibit 14.1   Criteria for Selection of Value Stocks

- Growth rate of sales exceeds 4%
- Growth rate of cash flow exceeds 7%
- Price growth persistence exceeds 50%
  OR
- Price stability exceeds 50%
- Return on capital exceeds 9%
- Value Index exceeds 18
- Price to cash flow ratio at or below the Value Line normal line

glamorous stocks with lower growth, we are willing to accept a growth rate of cash flow in the range of 8 to 12 percent. We look for boring, steady companies with good business practices. A typical company is Ecolab which provides cleaning chemicals to businesses. It has a growth rate of cash flow equal to 13.5 percent and a Rcap = 19%. Its Financial Strength is 1.1 (B++) and its Stock Price Stability is 95. Therefore, the sustainability is equal to

$$Sus = \frac{1.1 \times 95}{80} = 1.31$$

Therefore, the Power is

$$Power = 19 \times 13.5 \times 1.31 = 336$$

For a price of $37, the price to cash flow in mid-1999 was

$$PCF = \frac{\$37}{2.45} = 15.1$$

Therefore, the Value Index is equal to 22.3. This firm provided a return of 28.5 percent over the five-year period 1993–98 and should provide at least a return of 13% or 14% over the next five years.

## Selected Value Stocks

Table 14.1 provides a list of the performance measures for selected value stocks in diverse industries. Note that most of these companies are active in unglamorous industries and normally their PCF is less than 15. Champion Enterprises builds manufactured homes. Home buyers are responding to a shift to multiwide pre-made homes that offer quality, lower cost homes. Champion had $2.2 billion in sales in 1998 and sales are projected to grow at a rate of 12%. The 1999 PCF calculated with an average price as of April 1999 is a low 6.8. Thus, this value stock has a Value Index of 52.6.

Other examples of excellent value stocks are Central Newspapers, Ethan Allen, Graco, Maytag, and Ross Stores. Value stocks can always be found by looking for stocks with modest growth rates and low estimated price-to-cash-flow ratios.

## Setting a Strike Price for a Value Stock

One method of adding value stocks to your portfolio is to identify a company about which you gain knowledge of and confidence in their product or service by reading their reports and research reports. You then calculate their Power and set a minimum desired Value Index. Given this desired Value Index, the required price-to-cash-flow ratio can be calculated. Then, using the predicted cash flow, the necessary strike (purchase) price can be calculated. These steps are summarized in Exhibit 14.2.

As an example of this process, let's consider PepsiCo which provides beverage and snack food products. It spun off its restaurant businesses and is now in the growing

Table 14.1   Selected Value Stocks in Diverse Industries

| Name | Industry | Rcap (%) | G (%) | Sus | Power | Est PCF | Value Index | Debt Capital (%) | Profit Margin (%) | Growth Rate of Sales (%) | 1993–1998 Average Annual Return (%) |
|------|----------|------|------|------|------|------|------|------|------|------|------|
| Autodesk | Software | 21.5 | 17.5 | 0.6 | 226 | 7.7 | 29.3 | 0 | 15.7 | 14.0 | 12.1 |
| Alcoa | Aluminum | 12.0 | 11.5 | 0.98 | 135 | 8.4 | 16.1 | 33 | 5.6 | 8.5 | 18.8 |
| Avery Denison | Adhesives | 18.0 | 9.0 | 1.3 | 211 | 15.4 | 13.7 | 33 | 6.5 | 5.5 | 28.0 |
| Bemis | Packaging | 13.0 | 8.5 | 1.2 | 133 | 7.7 | 17.4 | 36 | 6.3 | 6.5 | 12.7 |
| Champion Enterprises | Manuf. Housing | 21.0 | 15.5 | 1.1 | 358 | 6.8 | 52.6 | 1 | 4.4 | 12.5 | 44.1 |
| Central News | Newspapers | 28.0 | 14.0 | 1.35 | 529 | 9.3 | 56.9 | 0 | 11.4 | 13.0 | 21.8 |
| Dana | Auto Parts | 16.0 | 9.5 | 1.0 | 152 | 5.5 | 27.6 | 46 | 5.6 | 5.5 | 9.6 |
| Dial | Soap, Food | 15.0 | 14.5 | 1.1 | 239 | 22.6 | 10.6 | 42 | 7.0 | 11.0 | 26.0* |
| Diebold | ATMs | 19.5 | 10.5 | 1.4 | 292 | 9.6 | 30.4 | 3 | 10.8 | 8.0 | 16.8 |
| Ecolab | Cleaning Chemicals | 19.0 | 13.5 | 1.3 | 335 | 15.1 | 22.2 | 23 | 8.4 | 11.0 | 28.5 |
| Ethan Allen | Home Furniture | 24.0 | 17.0 | 0.9 | 357 | 10.6 | 33.7 | 10 | 10.6 | 13.5 | 26.7 |
| Graco | Fluids Equipment | 34.0 | 14.5 | 1.0 | 493 | 6.9 | 71.4 | 62 | 11.5 | 8.5 | 28.1 |
| Illinois Tool | Metal Fabricating | 15.0 | 12.5 | 1.4 | 267 | 15.4 | 17.3 | 23 | 11.9 | 10.0 | 25.8 |
| Johnson Controls | Electrical Equip. | 13.5 | 12.5 | 1.4 | 240 | 6.7 | 35.8 | 38 | 2.6 | 10.0 | 20.0 |
| Kimberly Clark | Personal Care | 26.0 | 8.0 | 1.5 | 309 | 13.1 | 23.6 | 35 | 11.6 | 5.0 | 19.6 |
| Maytag | Appliances | 28.0 | 18.0 | 0.96 | 485 | 11.7 | 41.4 | 47 | 7.2 | 13.0 | 31.4 |
| Newell Rubbermaid | Housewares | 15.0 | 14.0 | 1.1 | 231 | 14.9 | 15.5 | 26 | 8.9 | 9.0 | 17.4 |
| PepsiCo | Beverages | 21.0 | 11.5 | 1.2 | 295 | 17.2 | 17.1 | 41 | 8.0 | 11.0 | 18.6 |
| Ross Stores | Clothing Retail | 27.0 | 18.0 | 0.83 | 401 | 10.6 | 37.8 | 7 | 6.0 | 13.5 | 44.8 |
| Sara Lee | Food Products | 27.5 | 10.0 | 1.4 | 380 | 11.8 | 32.2 | 53 | 5.7 | 5.0 | 20.3 |
| SBC Communicat. | Regional Telephone | 28.0 | 10.5 | 1.4 | 406 | 10.2 | 39.8 | 43 | 15.1 | 9.0 | 24.8 |
| Toll Bros. | Home Builders | 12.0 | 14.0 | 1.1 | 185 | 6.3 | 29.4 | 35 | 7.1 | 12.0 | 5.7 |
| Tosco | Petroleum Refining | 11.0 | 15.0 | 1.0 | 165 | 5.6 | 29.5 | 39 | 2.6 | 5.0 | 23.4 |
| Tommy Hilfiger | Clothes | 13.7 | 17.5 | 1.2 | 288 | 10.2 | 28.2 | 39 | 11.5 | 11.2 | 25.3 |
| Wausau-Mosinee | Paper | 12.5 | 14.5 | 0.7 | 125 | 5.8 | 21.6 | 24 | 6.7 | 12.0 | 0.0 |

*3 years
Prices as of April 1–15, 1999.

**Exhibit 14.2   The Strike Price for a Value Stock**

- Select a company with a good record and potential, and a moderate growth rate
- Calculate the power
- Set a minimum for the Value Index equal to 20 or more
- Calculate the required price-to-cash-flow ratio
- Using the predicted cash flow, calculate the strike price

beverage and snack food business. This company should provide significant increases in profits over the next five years. The Power is

$$Power = Rcap \times G \times Sus$$
$$= 21.0 \times 11.5 \times 1.22 = 295.$$

We will set a desired Value Index of 20. The equation for the Value Index is

$$Value\ Index = \frac{Power}{PCF}.$$

Solving for PCF we have

$$PCF = \frac{Power}{Desired\ Value\ Index}.$$

Then, with a desired Value Index = 20, we obtain

$$PCF = \frac{295}{20} = 14.75.$$

The predicted cash flow for 1999 for PepsiCo is $2.15. The desired strike price is then

$$Price = \left(\frac{Price}{Cash\ Flow}\right) \times Cash\ Flow$$
$$= 14.75 \times 2.15$$
$$= \$31.71.$$

The actual price in early 1999 ranged from a high of $41.50 in January to a low of $36.25 in late April. At the time of this writing (April 28, 1999) the price was $37.50. It is possible that PepsiCo would swing again into the $32 range where it would hit our strike price.

Another interesting example is Quantum Corporation which manufactures computer disk drives and tape storage devices. This industry is very competitive and Quantum is a good example of a value stock. Considering the competitiveness of this industry and the volatility of the stock price, we select a higher number for the desired Value Index and set the desired Value Index equal to 30. Calculating the Power we get

$$Power = Rcap \times G \times Sus$$
$$= 16.0 \times 15.5 \times 0.84$$
$$= 208.3.$$

Solving for the desired PCF we have

$$PCF = \frac{Power}{Desired\ Value\ Index}$$

$$= \frac{208.3}{30} = 6.94.$$

The predicted cash flow for 1999 is $2.50. Therefore, the strike price is

$$Price = PCF \times Cash\ Flow$$

$$= 6.94 \times 2.50 = \$17.35.$$

The actual price hit a low of $17 in April 1999 and the stock could have been purchased at that time.

## Socially Conscious Investing

Many investors seek to purchase the stock of companies that satisfy their social and environmental principles as well as provide an attractive return. Since we take the view that we are purchasing an ownership stake in a company when we purchase it, we need only add some criteria to our financial analysis. The investor will seek companies that tend to do well in financial measures and tend to do well in screens for progressive environmental practices, strong employee relations and benefits, and community involvement.

The idea of "values" investing gained momentum in the 1970s and 1980s. Investors seek financially strong companies that also meet their social and environmental goals. While mutual funds are available that use a screening technique to avoid certain industries, the investor is better served using an affirmative screening method. This method seeks companies that operate consistent with the investor's vision of a positive future. For example, many investors seek companies that demonstrate a high level of commitment to their workers, their communities and their environment.

A list of positive criteria for selecting companies is provided in Table 14.2. A list of eight stocks for socially conscious investors in given in Table 14.3. Two internet sites provide good information regarding social investment criteria: www.citizens-funds.com and www.socialinvest.org. However, investors should not overlook the companies' own internet sites and annual reports which usually incorporate corporate philanthropy and environment reports.

**Table 14.2   Social and Environmental Criteria**

- Generous and innovative giving to charities and local nonprofits
- Support for housing and education programs
- Family benefits such as child care
- Employment of the disabled
- Profit-sharing
- Pollution prevention
- Recycling and remanufacturing of waste and products
- Strong community relations
- Energy efficient operations and products

**Table 14.3    Eight Stocks for Socially Conscious Investors**

| Name | Symbol | Industry | Benefits and Values |
|---|---|---|---|
| AES Corp. | AES | Power generation | Environmentally competent, workplace safety |
| Cisco Systems | CSCO | Computers | Employee programs, flextime, school programs |
| Clorox | CLX | Domestic products | Environmentally competent, community involvement |
| IBM | IBM | Computers | Community programs, energy conservation |
| Herman Miller | MLHR | Furniture | Environmental and waste-reduction programs |
| Maytag | MYG | Elec. Appliances | Energy-efficient products |
| Robert Half | RHI | Temporary Staffing | Worker training and benefits |
| Solectron | SLR | Contract Manufacturer | Worker training, workplace safety |

# 15 Preparing for Market Declines

## This chapter covers

- Building a Portfolio
- Bear Markets
- Bubbles and Bears
- Selling Stocks
- Market Timing
- The Business Cycle

## Building a Portfolio

In previous chapters we have discussed how to purchase shares in an attractive business by purchasing stocks with a solid Power and a relatively high Value Index. We provided a method for selecting these stocks in four diversified portfolios that incorporate aggressive growth stocks in the Gazelle portfolio, solid and steady growth stocks in the Hare portfolio, slow-growing companies with solid dividends in the Tortoise portfolio, and out-of-favor stocks with good turnaround potential in the Contra portfolio.

We advocate purchasing the best possible portfolio and holding it for the long term. This method minimizes taxes and transaction costs and avoids trading on rumors and hunches. Using dollar-cost averaging tends to smooth out swings in the market and works efficiently for the average investor who periodically invests over a long period (typically 20 years).

A good rule for an investor is to create a solid investment plan, execute the plan over a long period, and build a reasonably diversified portfolio. Nevertheless, while we plan for the best outcome, we need to consider a strategy for negative surprises and market declines. In this chapter we consider several strategies for responding to unforeseen events that cause the overall stock market to decline significantly. A bull market is a sustained period of stock market growth as measured by a generally accepted market index such as the Dow Jones Industrial Average (DJIA) or the S&P 500 Index. While this selected market average usually continues to increase, it will decline from time to time. A **bear market** is a period of interruption of the upward trend of the market index market by a decline of 20% or greater from its latest high. There have been 22 bear markets in the 20th century. If the decline in the market index is 10% or greater but less than 20%, we call this decline a **market correction.**

## Bear Markets

During the period August 1896 through December 31, 1998 there have been 24 bear market periods of an average duration of about 12 months. The average duration of the bull markets during the same period was about 24 months. The average drop in a bear market was about 25% while the average gain during the bull markets was 75%. Thus, the investor stands to profit handsomely if the investor can continue to hold through a bear market drop of 25%.

The bear markets since 1950 are listed in Table 15.1. The longest bear market occurred in 1973–74 and lasted 22.8 months with a market decline of 45%. A bear market occurs when several trigger events occur. In 1973–74 there were five trigger events: 1) the Watergate scandal, 2) Vice President Spiro Agnew's resignation, 3) the Arab oil embargo, 4) the Israel-Syria war, and 5) President Richard Nixon's resignation. Furthermore, rising inflation persisted and gloom was rampant. The 1973–74 bear market was a very difficult period for all investors who feared that the nation might never emerge from all its problems.

The 1987 bear market was swift, short and shocking with a decline of 36%. This quick drop tested all buy-and-hold investors and many sold out at the bottom.

Bear markets start when share prices are high in relation to earnings and the Federal Reserve raises interest rates several times. Nevertheless, it is very difficult, if not impossible, to predict any bear market initiation date. The DJIA hit a high in July 1998 and then started to drop. Was this decline going to continue or was it just a normal correction? The DJIA then continued declining in August and September for a total decline of 16%, hitting a low in September 1998. This decline turned out to be a short-term correction and the market recovered all its losses by December 1998. In this case, the decline was a buying opportunity for investors.

Some investors even cheer when the market declines because they want to buy more shares at lower prices. If an investor loves ABC Corp. at $20, why not buy more at $15? The stock market takes dips, but if you own individual companies then it is

**Table 15.1    Bear Markets Since 1950**

| Period | Duration (months) | Decline (%) |
|---|---|---|
| 12/31/61 – 06/26/62 | 6.4 | 27.1 |
| 02/09/66 – 10/07/66 | 7.9 | 25.2 |
| 12/03/68 – 05/26/70 | 17.7 | 35.9 |
| 01/11/73 – 12/06/74 | 22.8 | 45.1 |
| 09/21/76 – 02/28/78 | 17.2 | 26.9 |
| 04/27/81 – 08/12/82 | 15.5 | 24.1 |
| 08/25/87 – 10/19/87 | 1.8 | 36.1 |
| 07/17/90 – 10/11/90 | 2.8 | 21.2 |
| **Average:** | **11.5** | **30.2** |

only necessary to know the value of your shares—not the whole market. Of course, the market usually takes all stocks down together (to a varying extent) and even great companies will experience the winds of change. The most likely event that will decrease the value of all stocks is one or two sequential increases in interest rates by the Federal Reserve. This will immediately increase the discount rate, r, that we use to calculate the appropriate value of our shares, thus reducing their intrinsic value. The other natural step towards reduction in value of a company's shares is a predicted decline in the growth rate of cash flow or earnings. A selloff in a market is not always a buying opportunity. If fundamentals decline, then the share price should also.

IBM was a powerful stock in 1973 and yet its price declined 48% in the two-year period from January 1, 1973, to December 31, 1974. The S&P 500 (without dividends) declined 42% in the same period. There were very few winners in that two-year market as interest rates climbed and business confidence crashed.

The PE ratio for a stock can be calculated from the discount model (Chapters 6 and 7) so that

$$PE = \sum_{n=1}^{N} x^n + \frac{1}{r} \qquad (15.1)$$

where $x = (1 + g)/(1 + r)$ and $g_p = 0$ is the growth rate in perpetuity. Then, when g and r are approximately equal we have $x = 1$ and therefore

$$PE = N + \frac{1}{r}. \qquad (15.2)$$

When a decline in confidence occurs, N decreases in the minds of the investors and as interest rates increase we have r increasing. Consider the simple case where N drops from 4 to 2 and r increases from 0.09 to 0.12. Remember, we are making the simplifying assumption that $g \approx r$. Then, the calculated PE ratio for this stock drops from 15 to 10, a 33% decline.

Another way to show the effect of a change in the growth rate and discount rate is to use the formula for a stock with a growth rate, g, in perpetuity. Then we have

$$PE = \frac{(1 + g)}{r - g} \approx \frac{1}{r - g}.$$

If bull market conditions exist and $g = 0.7$ and $r = 0.11$, we expect $PE = 25$. When interest rates go up so that $r = 0.12$ and g drops to 0.5, we have a calculated value for $PE = 14$.

## Bubbles and Bears

A physical bubble is a spherical, thin-walled object, often a result of an accumulation of gas. An **economic bubble** is a scheme or development that is insubstantial, groundless or ephemeral. Observers of economic history point to the 1637 Dutch Tulip bubble and the 1720 South Sea shares scheme to illustrate the idea of insubstantial or groundless valuation. When stock prices increase way beyond any reasonable valuation, observers label the stock market (or a sector) as a bubble. When Japanese investors vastly overpaid for real estate in 1990, many called it a real estate bubble.

Bubbles are simply extremes of a bull market. Often one or two sectors of the market will be very highly valued and a boom mentality prevails. A boom often grows to a manic phase which depends critically on the continuation of boom conditions. At the end of a boom, investor and television talk show comments often alternate between a discussion of a "new era" and paranoia about imagined future disasters.

Forecasters predict doom or boom and people write books about the "The Crash of '05." Forget these books and talk of new economic eras. Maintain a portfolio of powerful companies as well as other defensive stocks such as utilities or REITs as we outline in this book.

In the mania portion of a stock market, investors turn to speculation and trading. Then traders look to momentum as the driver of stock prices rather than fundamental measures like the discount model for earnings or cash flow. The power of a stock is summarized as:

$$\text{Power} = \text{Rcap} \times \text{G} \times \text{Sus}$$

which is analogous to the power of mechanical systems which is directly related to the product of force and velocity. However, the momentum of a physical object is mv where m = mass and v = velocity. The momentum within a stock market is the amount of money flowing into a stock or the volume of orders (or purchases) for a stock. Thus, stocks that investors want will be bid up by this desire to buy highfliers, only further feeding the mania.

A bull market bubble may expand until the frantic activity exhausts the participants. As the bubble grows, participants seem willing to take on increased risk as their self-confidence grows. The longer the bull market goes on, the less our memory of the last bear market. Excesses of inflation or corporate expansion and capacity can result in a bust in the bubble. Stay diversified and retain the defensive subportfolios: Tortoise and Contra.

As long as we stick to stocks with solid fundamentals as represented by Power and a Value Index greater than 20, we can usually avoid bubble stocks.

## Selling Stocks

Investors need a regular selling strategy so as to avoid panic selling at times of market depression. Exhibit 15.1 provides a list of reasons to sell some or all of the stock of a company. If the funds are required for an expense or the stock is greatly overpriced, you may wish to sell some shares in your portfolio. Keep the overall portfolio objectives in mind and sell some shares of your winners and your losers.

**Exhibit 15.1  Five Reasons to Sell Shares of a Stock**

- The money is needed for a planned expense
- The stock is greatly overpriced by the Power and Value Index method
- The industry in which the company operates is significantly deteriorating
- The fundamentals are deteriorating and the stock has disappointed recently
- A better opportunity is available

**Exhibit 15.2 Four Poor Reasons to Sell Shares of a Stock**

- Short-term news is negative
- The stock price rises to a new high
- The market appears to be overpriced and continues to rise in price
- You are convinced you can time the market

When the fundamentals of an industry or a specific company are declining and you cannot foresee a turnaround in the company shares, you may wish to sell. This sale may be particularly attractive if a solid, better opportunity is available.

Exhibit 15.2 gives four poor reasons to sell shares of a stock. If the near-term news of a company is negative, look again at the long-term reasons you used for buying this stock. Do they still hold?

If a stock price rises to a new high, it may be justified by the fundamentals. Furthermore, a small 5% swing in price is not a useful long-term signal for purchase or selling.

When the market appears to be overpriced and is continuing to rise it is not necessarily the right time to sell. The market may be discounting declining interest rates and confidence may be high, justifying the new price.

If you are convinced you can time the market, you should proceed to the next section.

## Market Timing

**Market timing** is the "timely" shifting of assets into or out of the market in an attempt to take advantage of market rallies while avoiding major declines. The essence of market timing in any investment is to buy low and sell high. While this is the dream of all investors, there is no realistic or guaranteed way to achieve it consistently. Fluctuations in the investment markets and their corresponding reflections in stock price changes are difficult to discern, and hunches about future market changes are unreliable.

Many people use one or more indicators to help them determine when to sell or buy a stock. These indicators are some type of measure of stock market or economic activity. However, it is unlikely that any investor can consistently use market timing effectively. In fact, investors who trade actively, trying to beat the market, probably will perform worse because of the high cost of commissions. The "modest" track record of professionals appears to confirm as much. Because the market is highly efficient, the theory states, the investor's best strategy is to buy and hold a portfolio that bears a degree of risk he or she is comfortable with.

Current market prices represent a consensus of all market participants. If this is so, then it is the sudden, unexpected arrival of new information that causes prices to change. But this new information cannot, by its very nature, be predicted in advance. And only those future events that fall within the realm of the "reasonably expected" are reflected in current market prices.

Market timing is, in practice, a hybrid of fundamental and technical methodologies, blending business analysis with technical projections. Market timers track such economic conditions as the trend and level of interest rates, the degree and direction of business activity, corporate profits, and industrial production.

Unfortunately, in the real world, market movements give dozens of signals, flashing buy, sell, and hold actions all at once. To further complicate the situation, institutions track each other's buy-and-sell actions as well as closely follow the same leading indicators. Thus, the major market players are all likely to react to the same information at about the same time.

Some market indicators may be meaningful to an investor in search of a sell or buy signal. However, even if the market gives you a signal, how does it fit the circumstances of your stock?

Value Line publishes a Technical rank for each stock in their Investment Survey. They also publish a Timeliness rank. Timeliness is based on the price momentum. Technical rank is based on trends of the stock price. The Timeliness and Technical ranks are highly correlated. Value Line statistical studies indicate that the Technical ranks perform best when growth stocks outperform those selected based on some value criteria and in markets favoring large capitalization issues over those with a small market cap.

The Value Line Timeliness and Technical indicators are short-term indicators of price momentum and can be used by traders. Since we are long-term investors we probably cannot profit from them.

As we showed earlier in this chapter, it is usually best to continue to hold your stock portfolio through a bear market. However, one indicator may be of use for determining the general thrust of valuation of the stock market. Interest rates directly influence the discount rate, r, for our discount model of equation 15.2. Thus, quite simply the intrinsic Value of our stocks declines as the Treasury bond rate increases. We propose a Market Attractiveness (MA) index as follows:

$$MA = PEDJ \times Y5T$$

where PEDJ is the current PE of the Dow Jones Industrial Average and Y5T is the yield on the five-year Treasury bond. Using Barrons Market Laboratory section, one can track MA weekly. Using data in the May 3, 1999 issue of *Barrons* we have

$$MA = 22.3 \times 5.10 = 113.7$$

from the Consensus Operating Earnings Estimates table and the five-year bond rate. The value of MA has held at 115 ± 15 over the past decade. Thus, as long as MA lies within the range of 100 to 130, we can state the market is fairly valued. In 1980, the PEDJ was 14 and the Y5T was 8.5. Then, the 1980 MA was 119. This indicator is interesting but is primarily of general interest value. It, like others, is of little use in timing the market. In any case, we do not buy the market, but rather we purchase ownership in great companies that we expect to weather through the occasional storm. In fact, weather forecasting is a fledgling science and often fails to accurately predict the weather. It is logical that the art of market predictions is of lesser value. We read the weather predictions and then proceed on our way. Similarly, we may read economic and market forecasts, but the best action is to proceed to buy and hold owner-ship in great companies.

## The Business Cycle

June is one of the most dangerous months for investing in stocks; the others are October, July, January, September, April, May, March, November, December, August, and February. That was Mark Twain's best stock market tip. Predicting whether a given share will rise or fall tomorrow, next week or next month is a hazardous business, best left to good luck, a blindfold and a pin. Investors can, however, be more confident about the performance of financial assets over the business cycle.

Financial markets, like the economy, move in cycles. Growth stocks, utility stocks, bonds, and cash are affected by the business cycle in different ways, so these are distinct periods to buy different sectors or types of assets. An idealized model of the over-all business cycle is shown in Figure 15.1. We start with Period 1, recession, where the growth in GDP is essentially zero or negative. The second period is the growth period and the third period occurs at the peak of the business cycle when the GDP stops growing. Eventually, the economy declines during the fourth period. The economy ex-perienced its last recession over the nine-month period June 1990 to March 1991. The economy has been in a growth period since 1991 to the time of this writing and some are predicting an entry into the 3rd period indicating a top in growth of GDP. At the end of the growth phase, the Federal Reserve senses rising inflation and raises interest rates. This act may lead to Phase 4 with declining rate of growth of GDP.

Exhibit 15.3 displays the characteristics of each segment of the business cycle and the best investment action for each segment.

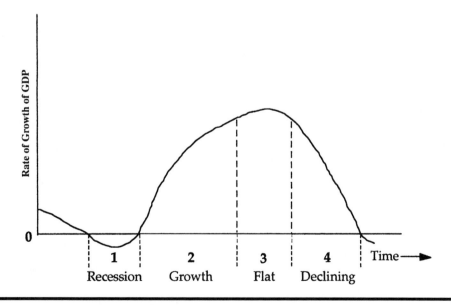

**Figure 15.1   The Business Cycle**

**Exhibit 15.3    Characteristics of the Business Cycle**

| Segment | Rate of Growth of GDP | Interest Rates | Inflation | Best Investment |
|---------|----------------------|----------------|-----------|-----------------|
| 1. Recession | Zero or Negative | Approaching a top | Topping | Aluminum companies Chemical Pharmaceuticals |
| 2. Growth | Increasing | Topping and then declining | Declining, flat later in this stage | Financial stocks Retail stores Technology |
| 3. Flat-Top | Flat | Declining, flat later in this stage | Starting to rise | Food stores Utilities REITs |
| 4. Declining | Declining | Rising | Increasing | Short-term bonds Cash Oil and gas companies |

# 16 Millennium Stocks

## This chapter covers

- A Portfolio of Millennium Stocks
- Investment Themes for the Millennium
- The Consumer Retail Sector
- The Metals, Energy, and Petroleum Sector
- The Leisure, Publications, Recreation, and Advertising Sector
- Industrial Supplies and Business Services Sector
- Manufacturing, Machinery, Tools, Electrical Equipment, and Aircraft Sector
- Banks, Insurance, and Financial Services
- Health Care, Medical Equipment, and Drugs
- Computer Hardware, Software, and Services
- Semiconductor Manufacturers and Equipment Makers
- Telecommunications Equipment and Services
- Utility Stocks
- REIT Stocks
- Contrarian Stocks

## A Portfolio of Millennium Stocks

In this chapter we build a portfolio of Millennium stocks suitable for most investors. In this section we identify the characteristics of our portfolio. In the next section we will describe the trends for the next millennium and select industries for investment consideration. In the final sections we actually proceed to identify 100 Millennium Stocks that will serve as candidates for your portfolio.

In earlier chapters we outlined the characteristics of a desirable portfolio. These characteristics are summarized in Exhibit 16.1. We seek to build a diversified portfolio of companies that consistently provide average annual returns that exceed 12%. These companies have a high Power and Value Index by providing a high return on capital and growth of cash flow and earnings. These companies are often members of growing industries that hold great potential over the next decade. The portfolio will be concentrated in less than 40 stocks so that they can be readily studied and understood. We will hold these stocks for the long term and avoid trading and timing

## Exhibit 16.1    Characteristics of a Powerful Portfolio

- A diversified portfolio of companies that consistently provide excellent returns.
- A reasonable number of stocks, less than 40.
- Great, powerful companies in attractive, growing industries.
- Hold for the long term; avoid trading and timing.
- Select companies with high return on capital and growth of cash flow.
- Buy stocks when their price to cash flow is near the Value Line normal PCF.
- Follow the companies in your portfolio.
- Learn to know your companies.
- Revise the portfolio towards the winners.

strategies. We purchase a stock when their price to cash flow (PCF) is near or at the Value Line normal PCF for that stock. Over the long term, we sell the nonperformers and build the portfolio toward the winners.

The idea is to build a portfolio with great companies so that you can build your confidence in them. As volatility in the market occurs, you use the market declines to add to your holdings of these great companies. The average annual turnover of your portfolio will be less than 10% and you will minimize taxes and trading costs. We want companies with a high probability of achieving the expected return over a five-year period. When the probability is high, then the risk of loss is low. We are not necessarily seeking low volatility. Rather we seek a high probability that we achieve our expected returns [Hagstrom, 1999].

A suggested set of portfolios are provided in Table 16.1. You may wish to start with a conservative portfolio and shift over time towards the moderate or the aggressive portfolio. The maximum size of the portfolio is less than 40 stocks so that we can monitor the companies that we own.

We seek to build a portfolio of powerful stocks using the stock selection criteria given in Exhibit 16.2. The companies meeting this criteria will normally provide an investor with a very attractive return over the long term.

### Table 16.1    Suggested Portfolios

|  | Conservative | | Moderate | | Aggressive | |
|---|---|---|---|---|---|---|
|  | Number | Percent (%) | Number | Percent (%) | Number | Percent (%) |
| Contras | 2 | 8.3 | 2 | 8.3 | 3 | 10.7 |
| Tortoises | 10 | 41.7 | 10 | 35.7 | 9 | 28.1 |
| Hares | 10 | 41.7 | 12 | 42.9 | 12 | 37.5 |
| Gazelles | 2 | 8.3 | 3 | 10.7 | 8 | 25.0 |
| **Total:** | 24 | 100% | 28 | 100% | 32 | 100% |

## Exhibit 16.2    Stock Selection Objectives

- Power exceeds 250.
- Value index exceeds 18.
- High return and high probability of attaining that return.
- Industry with sales growth exceeding 10%.
- Top first or second stocks with the best performance characteristics in their industry.
- High confidence in the management of the company.

**Table 16.2   Confidence in Achieving an Expected Return**

| Confidence | Very Low | Low | Medium | High | Very High |
|---|---|---|---|---|---|
| Rank | 1 | 2 | 3 | 4 | 5 |
| Probability (P) | 0.05 | 0.30 | 0.50 | 0.80 | 0.99 |

We will estimate an expected total return, ER, for a company using the Value Line Survey and other information sources. Then, based on an estimate of our confidence in the company we estimate a probability, P, of achieving the expected return. Then, the probable return, PR, is

$$PR = ER \times P.$$

The confidence ratings we use are shown in Table 16.2. Of course, we would not normally purchase any company's stocks that have a low or very low confidence rating. If a stock has an expected return of ER = 20% and a confidence rank of 4 (high) we determine the probable return as

$$PR = 20\% \times 0.8 = 16\%.$$

Naturally, we plan to build our portfolio over time using stocks with a confidence rank of 4 or 5. Most investors will start their portfolio with five to 10 stocks and build over the years to a portfolio of 20 to 30 stocks. As the high-return stocks emerge and prove themselves, they will become a large part of our portfolio. We do not try to rebalance the portfolio, but rather let the natural organic growth occur. Thus, after ten years we may find that 10 winning stocks have grown to account for 90% of our portfolio.

## Investment Themes for the Millennium

Many business and social themes are emerging to provide important business and investment opportunities. Exhibit 16.3 lists 12 important themes for the next decade. The baby boomers and the wave of immigrants to the U.S. will support consumer spending for home furnishing, appliances, and home remodeling projects. A strong trend towards restaurants with specialty products such as coffee, tea, or clothing is strengthening. The use of the internet for direct selling of products and services is growing. Increasing number of stores are specializing in higher value-added products and services. As boomers age they spend as well as save for retirement through mutual funds and securities brokers.

Technology has a large effect on productivity. U.S. capital investment, in the past 10 years (1989–99), has doubled from approximately 7% of Gross Domestic Product (GDP) to 13%. This surge in corporate investment has accounted for 25% of GDP growth. We are in the midst of transforming the distribution of goods and services and knowledge through the internet. Productivity, as a result, has increased from the 1 % of the 1980s to over 2% per year in the 1990s. Information technology has accounted for much of the productivity gain. This technology is well suited for the transactional activity which accounts for up to 40% of the economic activity in the industrialized nations. Most industries are spending heavily to boost productivity and expand

**Exhibit 16.3    Business and Social Themes for the New Millennium**

- Strong consumer spending on home furnishings, appliances, and remodeling projects.
- Specialty restaurants and coffeehouses.
- Direct producer-to-customer selling.
- Customized products.
- Savings flowing into stocks and mutual funds.
- Increasing financial transactions.
- Expansion of the internet and intranets.
- Growth of telecommunication services and equipment.
- Outsourcing of manufacturing and business services.
- Expansion of leisure and recreation and worldwide brands.
- Growth of global trade.
- Growth of travel and transportation of goods.
- Semiconductor design and manufacturing.
- Increased need for health services, equipment, and drugs.

markets. Nevertheless, it will take several years to work off the excess capacity in steel, metals, autos, and commodity products. Thus, we will deemphasize commodities and metals.

The sections that follow will consider broad industry classifications and describe the companies that are attractive. We use a table to provide the projected return on capital, growth rate of cash flow, Power and Value Index for each stock. We also indicate the expected growth rate of sales, expected return, confidence rank for the expected return, and the calculated probable return. The Value Index is calculated using the Value Line normal price-to-cash-flow ratio. We assume the purchaser can attain the stock at a time when the PCF approaches the normal Value Line number. For a long-term holder, it is not critical to wait for the PCF to equal the normal Value Line PCF. We only list companies that we have selected for membership in the 100 Millennium Stocks. We tested other stocks, but this list contains only the best.

# The Consumer Retail Sector

Here we consider the expected increase in consumer spending and home furnishings, apparel, food services, home improvements, and home products. This is a very large sector with 16 Millennium Stocks as listed in Table 16.3. The moderate growth, Hare, stocks such as Coca-Cola, Clorox, and Walgreen provide consistent returns in the 15% to 20% range. The high growth, Gazelle, stocks such as Gap, Starbucks, and Home Depot provide probable returns in the range from 20 to 30 percent.

Albertson's is the second largest food retailer in the U.S. Albertson's is known for its excellent operating skills and should be able to deliver consistent 15% annual returns.

Bed, Bath, & Beyond is a leading retailer of home furnishings and home products such as lamps, towels, and cookware. BBBY currently operates about 200 stores in 35 states. The growth rate of sales has been about 30% for the past five years.

Starbucks opens one new store each day. It has an excellent brand. They are moving overseas and into supermarkets. They are a powerful growth machine and may hold their growth for at least another five years.

Home Depot is the world's largest home-improvement retailer with about 800 stores worldwide. It aims to have 1,600 stores by 2002. It has a great business model and should grow sales at 20% annually over the next five years. This is a core holding for any portfolio.

Gap is a clothing retailer with 2,000 stores and a strong management. It has become a trendsetter for men and women. It should have an excellent potential as it moves overseas with its excellent brands.

Jones Apparel is the largest multibrand fashion company, selling its products through more than 6,200 locations. The company is growing sales at 15% per year.

## The Metals, Energy, and Petroleum Sector

The metals, energy and petroleum sector is dependent upon the price of oil and natural gas and metals. We select only four stocks as candidates for our Millennium portfolio as shown in Table 16.4. If energy and metals prices rise significantly, then other stocks may warrant consideration. All four stocks are moderately growing companies with sound management and reasonable prospects. Schlumberger is the world's leader in wellsite drilling services for the oil and gas industry. It should be able to increase its growth of cash flow over the next five to ten years. It is an excellent long-term holding.

Alcoa has fashioned itself as the efficient aluminum producer and has a year's lead over its competitors. Productivity is key to Alcoa's success.

## The Leisure, Publications, Recreation, and Advertising Sector

As the potential time for leisure, travel and entertainment increases people spend more for good experiences. Table 16.5 lists five companies that are active in the newspaper, travel, and recreation sector. The hotel industry is subject to oversupply and recession. However, Marriott International appears to be well-positioned to weather the hotel business cycles and provide a solid, moderate return over the long run. Carnival is a cruise operator with good growth characteristics. Carnival is focused towards the younger market and is expected to experience a 14% growth in sales over the next five years.

Harley-Davidson is the largest U.S. maker of heavyweight motorcycles and is famous for its long waiting lists of customers. Their production capacity is balanced and the company should be able to grow its sales about 15% per year.

Interpublic Group is one of the two largest advertising companies in the world. About one third of its growth is from acquisitions. About 50% of its revenues are from international markets.

## Industrial Supplies and Business Services Sector

The industrial supplies and business services sector is an important part of the U.S. economy that enables businesses to meet the demands of their customers and contract out noncore functions such as shipping, payroll, and staffing. This business-to-business sector also has the companies that supply critical supplies such as furniture, office machines, copiers, and cleaning materials. Table 16.6 lists nine companies that excel in the industrial services and supplies sector.

**Table 16.3 The Consumer Sector (Projected 1999–2004)**

| Name | Industry | Rcap (%) | G (%) | Power | Value Line PCF | Value Index | Growth Rate of Sales (%) | Expected Return (%) | Confidence Rank (%) | Probable Return (%) |
|---|---|---|---|---|---|---|---|---|---|---|
| Jones | Apparel | 20 | 20 | 374 | 13 | 28.8 | 16.5 | 20 | 5 | 20 |
| Gap | Apparel | 35 | 26 | 880 | 14 | 63.0 | 25.5 | 25 | 5 | 25 |
| Coca-Cola | Beverage | 49 | 9 | 556 | 24 | 23.1 | 7.0 | 18 | 4 | 14 |
| Sara Lee | Food | 27.5 | 10 | 304 | 12 | 25.3 | 5.0 | 15 | 5 | 15 |
| Starbucks | Food | 15.0 | 25.0 | 413 | 15 | 27.5 | 23.5 | 30 | 4 | 24 |
| Albertsons | Food store | 16.0 | 11.0 | 183 | 12.5 | 14.6 | 9.5 | 15 | 5 | 15 |
| Walgreen | Drug store | 17.0 | 14.5 | 288 | 17.0 | 17.0 | 12.0 | 22 | 5 | 22 |
| Ethan Allen | Furniture | 23.5 | 17.0 | 280 | 10.5 | 26.6 | 13.5 | 25 | 4 | 20 |
| Home Depot | Hardware | 19.0 | 23.0 | 489 | 21.0 | 23.3 | 21.0 | 25 | 5 | 25 |
| Bed, Bath & Beyond | Furnishings | 21.0 | 25.5 | 578 | 21.5 | 26.9 | 26.0 | 30 | 5 | 30 |
| Clorox | Home items | 22.0 | 15.0 | 439 | 12.5 | 35.1 | 11.0 | 20 | 5 | 20 |
| Avon | Cosmetics | 60.0 | 15.0 | 941 | 17.0 | 55.3 | 9.0 | 20 | 5 | 20 |
| Kimberly-Clark | Paper | 28.0 | 8.0 | 251 | 12.0 | 20.9 | 5.0 | 15 | 5 | 15 |
| Dollar General | Discount | 22 | 20.5 | 514 | 15 | 34.3 | 21.0 | 30 | 4 | 24 |
| WalMart | Discount | 19.5 | 16.5 | 292 | 16 | 18.3 | 16.0 | 20 | 4 | 16 |
| Maytag | Appliance | 28.0 | 18.0 | 416 | 10 | 41.6 | 13.0 | 18 | 4 | 14 |

**Table 16.4 The Metals, Energy and Petroleum Sector (Projected 1999–2004)**

| Name | Industry | Rcap (%) | G (%) | Power | Value Line PCF | Value Index | Growth Rate of Sales (%) | Expected Return (%) | Confidence Rank (%) | Probable Return (%) |
|---|---|---|---|---|---|---|---|---|---|---|
| Schlumberger | Oil | 16.5 | 11.0 | 203 | 12.0 | 16.9 | 9.0 | 15.0 | 4 | 12 |
| Exxon | Oil | 15.5 | 7.0 | 144 | 10.0 | 14.4 | 4.0 | 14.0 | 5 | 14 |
| El Paso Energy | Natural gas | 9.0 | 17.0 | 122 | 8.0 | 15.3 | 14.5 | 18.0 | 4 | 14 |
| Alcoa | Aluminum | 15.5 | 13.0 | 183 | 7.0 | 26.2 | 7.5 | 13.0 | 4 | 10 |

**Table 16.5 The Leisure, Publications, and Advertising Sector (Projected 1999–2004)**

| Name | Industry | Rcap (%) | G (%) | Power | Value Line PCF | Value Index | Growth Rate of Sales (%) | Expected Return (%) | Confidence Rank (%) | Probable Return (%) |
|---|---|---|---|---|---|---|---|---|---|---|
| Interpublic | Advertising | 18.0 | 15.0 | 333 | 13.5 | 24.7 | 12.0 | 20.0 | 5 | 20 |
| Marriott | Hotels | 13.0 | 15.5 | 181 | 14.0 | 13.0 | 15.0 | 16.0 | 4 | 13 |
| Harley-Davidson | Motorcycles | 17.0 | 18.5 | 329 | 12.5 | 26.3 | 15.5 | 17.0 | 5 | 17 |
| Carnival | Cruise Line | 15.5 | 16.5 | 267 | 14.0 | 19.1 | 14.0 | 17.0 | 4 | 14 |
| Central News | Newspaper | 30 | 14 | 454 | 11.0 | 41.2 | 13.0 | 15.0 | 5 | 15 |

**Table 16.6 Industrial Supplies and Business Services Sector (Projected 1999–2004)**

| Name | Industry | Rcap (%) | G (%) | Power | Value Line PCF | Value Index | Growth Rate of Sales (%) | Expected Return (%) | Confidence Rank (%) | Probable Return (%) |
|---|---|---|---|---|---|---|---|---|---|---|
| FDX | Package delivery | 12 | 11.5 | 121 | 5.5 | 22.1 | 7.0 | 16 | 4 | 13 |
| Ecolab | Cleaning supplies | 19 | 13.5 | 268 | 10.0 | 26.8 | 11.0 | 16 | 5 | 16 |
| Staples | Office supplies | 18 | 25 | 540 | 19.0 | 28.4 | 18.5 | 30 | 5 | 30 |
| Xerox | Copiers & printers | 15.5 | 12 | 218 | 11.5 | 19 | 8.0 | 20 | 5 | 20 |
| Pitney Bowes | Office machines | 28 | 10.5 | 282 | 12.5 | 22.6 | 9.5 | 25 | 4 | 20 |
| Equifax | Credit reports | 23 | 14 | 377 | 14 | 26.9 | 8.0 | 20 | 5 | 20 |
| Robert Half | Staffing | 20 | 29.5 | 566 | 21 | 27.0 | 28.0 | 25 | 4 | 20 |
| Paychex | Payroll | 30 | 24 | 752 | 28.5 | 26.4 | 18.5 | 30 | 5 | 30 |
| Herman Miller | Furniture | 24 | 14.5 | 292 | 10 | 29.2 | 11.5 | 18 | 4 | 15 |

Ecolab supplies cleaning materials for offices, restaurants, and manufacturers. The company's sales have risen over 10% annually for the past seven years and it is the leader in its industry.

Staples provides office supplies and Herman Miller provides office furniture. Both of these firms are the leaders in their respective industries.

Xerox is set to be a key supplier of copiers, printers, and fax machines for the digital documents and digital office that is part of the digital network.

Paychex provides computerized payroll, accounting, and tax return services to more than 300,000 small and medium sized businesses nationwide. The growth of revenues has exceeded 16% annually over the past decade and that rate of growth is expected to continue over the next five years.

## Manufacturing, Machinery, Tools, Electrical Equipment, and Aircraft Sector

Table 16.7 lists 10 leading companies in this broadly defined sector focused on manufacturing in various fields. We also include the leading environmental management company in this broad classification. The manufacturing sector has experienced a solid uptrend over the past decade. In addition, as the economic problems overseas are resolved, this sector should experience solid, newly restored growth in international sales.

The production line that makes HP Inkjet printers in Newark, CA, is impressive and well-organized. This factory is owned and operated by Solectron, one of the leaders in the contract manufacture business. Solectron provides efficient services and has experienced growth of revenues exceeding 20%. Solectron is the world's largest contract electronics maker.

Solectron is a powerful behind-the-scenes force in electronics, assembling products and components for companies such as IBM, HP, and Cisco. Solectron is rapidly expanding around the world.

We expect Boeing to improve their operating margins as aircraft sales increase. This sector also includes the powerhouse manufacturing companies such General Electric, Allied Signal, Carlisle, Tyco, and Johnson Controls. All these firms are able to grow earnings at about 15% annually and thus provide a sustained return of 15% to 18% for shareholders.

We also include Illinois Tool Works (ITW) that can provide a steady 18% return. This company makes welding equipment, components, fasteners, and tools for many industries. It has experienced uninterrupted growth of cash flow over the 15 years.

Waste Management is the largest solid-waste disposal company in North America. This company supplies a critical service and should grow as the need for this service grows.

## Banks, Insurance, and Financial Services

With a healthy U.S. economy, the demand for loans, bank services, insurance and financial services should remain strong. Banks with overseas operations will find restored growth as the Asian and Latin American financial crises abate. Many banks and financial companies are merging and cost reductions are steadily occurring.

**Table 16.7  Manufacturing, Machinery, Tools, Electrical Equipment, Aircraft (Projected 1999–2004)**

| Name | Industry | Rcap (%) | G (%) | Power | Value Line PCF | Value Index | Growth Rate of Sales (%) | Expected Return (%) | Confidence Rank (%) | Probable Return (%) |
|---|---|---|---|---|---|---|---|---|---|---|
| Illinois Tool | Tools | 15 | 12.5 | 214 | 15 | 14.3 | 10.0 | 18 | 5 | 18 |
| Solectron | Contract manufacturer | 18.5 | 28 | 570 | 14 | 40.7 | 32.0 | 30 | 5 | 30 |
| Boeing | Aircraft | 21 | 17 | 300 | 12 | 25.0 | 11.5 | 16 | 4 | 13 |
| Johnson Controls | Auto parts | 14 | 12.5 | 200 | 5.5 | 36.4 | 10.0 | 15 | 5 | 15 |
| Carlisle | Plastics | 16.5 | 18 | 321 | 10 | 32.1 | 17.0 | 18 | 4 | 14 |
| Tyco | Diversified | 24 | 24 | 484 | 12 | 40.3 | 12.5 | 16 | 5 | 16 |
| General Electric | Electrical | 27 | 13 | 491 | 18 | 27.3 | 5.5 | 22 | 5 | 22 |
| Allied Signal | Aerospace | 18.5 | 10.5 | 272 | 13 | 20.9 | 6.5 | 15 | 5 | 15 |
| Champion Enterprises | Manufactured Homes | 18 | 15.5 | 251 | 11 | 22.8 | 12.5 | 20 | 5 | 20 |
| Waste Management | Environment | 14.5 | 19.5 | 283 | 12 | 23.6 | 16.5 | 18 | 4 | 15 |

Table 16.8 lists 13 companies active in banking, insurance, and financial services. See Chapter 9 for a discussion of this sector. With steady, low interest rates these firms all may perform well over the next five years.

State Street Corporation (SST) is an excellent example of a leader in the provision of financial services. It provides a range of trust services to mutual fund companies and U.S. pension funds. With $5.0 trillion in assets under custody and $525 billion under management, State Street is one of the world's leading specialists in serving institutional investors. With the expected increase in retirement funds in the U.S. and overseas, State Street should continue to grow its revenues and earnings at a compound annual rate of 16%. SST has invested heavily in technology to ensure the company offers innovative, reliable products to its institutional clients. Many of State Street's competitors exited the field because they did not have the technology to survive.

This sector includes four great banks—First Union, Mellon, Firstar, and U.S. Bancorp—that should be able to return about 20% annually. We include one thrift, Washington Mutual (WM), which should benefit from favorable economic conditions. WM is the largest thrift in the U.S. with over 2,000 offices in 40 states. This thrift makes loans and offers mortgages as well as other services.

MBNA and Providian are consumer lenders through credit cards and other means. Both firms should be able to grow earnings at 25% per year or better.

## Health Care, Medical Equipment, and Drugs

Health care, medical equipment, and drugs account for about 14% of the U.S. Gross Domestic Product. The industries within the health-care sector that have provided the highest returns are drugs and medical technology. Hospitals and Health Maintenance Organizations (HMOs) have been less attractive investments. Table 16.9 lists nine health-care firms in the drugs, equipment, and supplies industries. See Chapter 8 for an earlier discussion of the pharmaceutical industries.

Abbott Laboratories (ABT) and Johnson & Johnson (JNJ) manufacture a broad line of products for health care. Both have provided steady earnings and revenue growth in the 10–12% range for many years. We expect to see growth for both firms at the 12% level, and their strengths in pharmaceuticals should sustain their performance.

Pfizer is a drug company with great products and strong marketing. Pfizer concentrates on patent-protected drugs that produce a healthy 32% operating margin on its sales of $13 billion. Pfizer provided an average annual return of 51% for the five-year period 1993–98. We expect a return of 25% per year for the period 1999–2004.

Medtronic and Guidant are leading providers of medical equipment such as heart pacemakers, stents, and other technology. Both firms have excellent records of innovation of new products. We expect returns exceeding 25% for both firms.

## Computer Hardware, Software, and Services

The computer hardware, software, and services sector contained some very powerful winners over the past decade. Examples of great hardware companies are Dell, Cisco,

**Table 16.8  Banks, Insurance, and Financial Services (Projected 1999–2004).**

| Name | Industry | RoE (%) | G* (%) | Power | Value Line PE | Value Index | Return on Assets (%) | Expected Return (%) | Confidence Rank (%) | Probable Return (%) |
|------|----------|---------|--------|-------|---------------|-------------|----------------------|---------------------|---------------------|---------------------|
| First Union | Bank | 24 | 14 | 363 | 12 | 30.2 | 2.9 | 18 | 5 | 18 |
| Mellon | Bank | 24 | 17 | 427 | 18 | 23.7 | 2.5 | 25 | 5 | 25 |
| Firstar | Bank | 19.5 | 18 | 367 | 16.5 | 22.2 | 2.3 | 20 | 5 | 20 |
| US Bancorp | Bank | 27 | 15 | 405 | 13 | 31.2 | 2.9 | 25 | 5 | 25 |
| Washington Mutual | Thrift | 18 | 21 | 386 | 14 | 27.5 | 1.3 | 22 | 5 | 22 |
| MBNA | Financial | 32 | 29.5 | 1,038 | 17.5 | 59.3 | 1.3 | 28 | 5 | 28 |
| Providian | Financial | 23.5 | 35.5 | 834 | 24 | 34.8 | 5.5 | 35 | 5 | 35 |
| American International | Insurance | 13 | 15.5 | 262 | 19 | 13.8 | 1.8 | 25 | 5 | 25 |
| Schwab | Securities | 25 | 22 | 605 | 32 | 18.9 | 2.1 | 25 | 5 | 25 |
| T. R. Price | Mutual funds | 28.5 | 17.5 | 599 | 20 | 29.9 | 1.2 | 30 | 5 | 30 |
| State Street | Financial | 18.5 | 21.5 | 382 | 15 | 25.5 | 1.2 | 25 | 5 | 25 |
| Capital One | Financial | 19 | 27.0 | 513 | 20 | 25.7 | 3.0 | 35 | 4 | 28 |
| Fannie Mae | Financial | 20 | 15.5 | 403 | 16 | 25.2 | 0.9 | 25 | 5 | 25 |

*G = Growth Rate of Earnings

**Table 16.9 Health Care, Medical Equipment, and Drugs (Projected 1999–2004)**

| Name | Industry | Rcap (%) | G (%) | Power | Value Line PCF | Value Index | Growth Rate of Sales (%) | Expected Return (%) | Confidence Rank (%) | Probable Return (%) |
|---|---|---|---|---|---|---|---|---|---|---|
| Pfizer | Drugs | 38.0 | 17.5 | 791 | 19 | 41.7 | 14.0 | 25 | 5 | 25 |
| Schering Plough | Drugs | 40 | 16.0 | 806 | 19 | 42.4 | 14.0 | 25 | 5 | 25 |
| Abbott | Products | 28.5 | 9.5 | 322 | 16 | 20.1 | 7.0 | 18 | 5 | 18 |
| Johnson & Johnson | Products | 29 | 12 | 438 | 14.5 | 30.2 | 13.0 | 20 | 5 | 20 |
| Medtronic | Equipment | 15 | 14 | 273 | 18 | 15.2 | 12.0 | 25 | 4 | 20 |
| Guidant | Equipment | 30 | 23 | 759 | 20 | 38.0 | 20.0 | 30 | 5 | 30 |

**Table 16.10 Computer Hardware, Software, and Services (Projected 1999–2004)**

| Name | Industry | Rcap (%) | G (%) | Power | Value Line PCF | Value Index | Growth Rate of Sales (%) | Expected Return (%) | Confidence Rank (%) | Probable Return (%) |
|---|---|---|---|---|---|---|---|---|---|---|
| Dell | Hardware | 47 | 32 | 2,000 | 31 | 64.5 | 27.0 | 35 | 5 | 35 |
| Cisco | Hardware | 27 | 25.5 | 964 | 25.5 | 37.8 | 26.0 | 35 | 5 | 35 |
| IBM | Hardware & Software | 26.5 | 13 | 434 | 10.0 | 43.4 | 13.0 | 30 | 5 | 30 |
| EMC | Storage | 25.5 | 32.5 | 1,102 | 30 | 36.7 | 29.5 | 35 | 5 | 35 |
| Sun | Hardware & Software | 25.0 | 17.5 | 459 | 16.5 | 27.8 | 14.5 | 30 | 5 | 30 |
| AOL | Software | 37.5 | 50 | 1,875 | 45 | 41.6 | 36.0 | 45 | 4 | 36 |
| Computer Associates | Software | 29 | 15 | 452 | 11 | 41.1 | 16.0 | 30 | 5 | 30 |
| Microsoft | Software | 25 | 27 | 898 | 30 | 29.9 | 23.5 | 35 | 5 | 35 |
| Oracle | Software | 34 | 22.5 | 780 | 25 | 31.2 | 26.5 | 35 | 4 | 28 |
| Automatic Data | Services | 21 | 14.5 | 405 | 15.5 | 26.1 | 13.0 | 18 | 5 | 18 |

EMC, and Sun. During the 10-year period, 1989–98, the average annual return of the aforementioned companies ranged from a low of 35% for Sun Microsystems to a high of 80% for Dell. This growth of earnings and associated return was largely a result of the explosion of the purchases of personal computers. In the next decade we will see lower-priced personal computers (PC) as well as other personal computing devices such as handheld computers, network computers, personal digital assistants, laptop computers, and television-top boxes. Also, the complicated PC may be replaced by simpler computing appliances. PC household penetration is at 50% in the U.S. and may mature at that level as simpler devices become available for the home. Worldwide there are about 200 million PCs. Over the next five years, growth worldwide should bring the PCs worldwide to at least 800 million. Sun, IBM, Dell, and EMC should grow with the PC expansion and the internet usage of computers.

Table 16.10 shows the 10 leaders in the computer industry. Dell is a direct marketer of PCs and larger network computers. Dell receives an order either via telephone or the internet and then builds the computer on the spot. That means the Texas-based company can keep inventory low while providing custom PCs.

Computer makers make larger margins when they sell workstations and network servers. IBM, Compaq, Sun, and Cisco sell these larger computers.

Software and services are higher-margin businesses and provide solid returns to companies like Microsoft, Computer Associates, Oracle, and Automatic Data Processing. EMC manufactures storage devices for network computer systems. EMC offers storage systems for computers and vast electronic "warehouses" of operating data held on mainframes or servers. What differentiates EMC products is its software, which coordinates the movement of data. EMC provided an annual return of 69% during the ten years 1989–98.

Cisco is the market leader in computers for data networks. 1998 sales were up 31% to $8.5 billion and income of $1.3 billion for a profit margin of 15%.

Microsoft and Oracle are powerhouse software companies that race each other to the leading edge. Both firms may return 30% or better over the next five years. Microsoft returned 58% per year over the 10-year period 1989–98 while Oracle provided a 40% return over the same period.

## Semiconductor Manufacturers and Equipment Makers

Table 16.11 lists the five leaders in the semiconductor manufacturer and equipment makers sector. The manufacturers of integrated circuits (ICs) are Intel, Linear Technology, Vitesse, and Xilinx. Applied Materials provides the process equipment to the manufacturers. The IC manufacturers experience growth cycles as companies worldwide periodically overbuild manufacturing capacity. The growth of the internet and telecommunications is raising demand for ICs. Over the next five years, the projected growth in demand for IC sales should exceed 15%. As the excess capacity is absorbed, the market for Applied Materials equipment should grow at 15% per year. This can lead to an expected return for Applied Materials of 25%.

The convergence of computers, communications and the internet should push demand for ICs. Intel is the world's largest producer of microprocessor ICs for personal

**Table 16.11   Semiconductor Manufacturers and Equipment Makers (Projected 1999–2004)**

| Name | Industry | Rcap (%) | G (%) | Power | Value Line PCF | Value Index | Growth Rate of Sales (%) | Expected Return (%) | Confidence Rank (%) | Probable Return (%) |
|---|---|---|---|---|---|---|---|---|---|---|
| Applied Materials | Equipment | 18.5 | 17 | 359 | 14.5 | 24.7 | 15.0 | 25 | 5 | 25 |
| Intel | Integ. Ckt. | 26.5 | 16 | 594 | 13 | 45.7 | 15.5 | 30 | 5 | 30 |
| Linear Technology | Integ. Ckt. | 23 | 20.5 | 538 | 20 | 26.9 | 20.0 | 30 | 5 | 30 |
| Vitesse | Integ. Ckt. | 20.5 | 28 | 568 | 25 | 22.7 | 26.0 | 30 | 5 | 30 |
| Xilinx | Integ. Ckt. | 20 | 16 | 317 | 15.5 | 20.4 | 14.0 | 25 | 5 | 25 |

**Table 16.12   Telecommunications Equipment and Services (Projected 1999–2004)**

| Name | Industry | Rcap (%) | G (%) | Power | Value Line PCF | Value Index | Growth Rate of Sales (%) | Expected Return (%) | Confidence Rank (%) | Probable Return (%) |
|---|---|---|---|---|---|---|---|---|---|---|
| Lucent | Equipment | 21.0 | 19 | 519 | 25 | 20.7 | 12.5 | 30 | 5 | 30 |
| Nokia | Equipment | 25.5 | 25.5 | 663 | 18 | 36.8 | 26.0 | 30 | 5 | 30 |
| Tellabs | Equipment | 18 | 24 | 518 | 20 | 25.9 | 21.0 | 35 | 5 | 35 |
| JDS Uniphase | Equipment | 20 | 40 | 720 | 25 | 28.8 | 35.0 | 35 | 5 | 35 |
| AT&T | Services | 17 | 14 | 270 | 10 | 27.0 | 4.0 | 18 | 5 | 18 |
| MCI-Worldcom | Services | 10 | 20 | 240 | 14 | 17.1 | 20.0 | 20 | 5 | 20 |
| SBC | Services | 15 | 10.5 | 175 | 10 | 17.5 | 9.0 | 18 | 5 | 18 |
| Qualcomm | Equipment | 15.5 | 24 | 518 | 20 | 25.9 | 32.5 | 30 | 4 | 24 |

computers with about 80% of the market. It is also pursuing the market for ICs for the data and video communications market.

Vitesse Semiconductor Corporation is a leading supplier of ICs for high performance telecommunications equipment. Vitesse stands to have revenue growth of 25% annually and may return 30% per year.

Linear Technology is a leader in the manufacture of high-performance integrated circuits for the transformation or amplification of analog signals (as contrasted to digital signals). Linear Technology provided a return of 46% over the 10-year period 1989–98. We expect a 30% return over the period 1999–2004.

## Telecommunications Equipment and Services

Table 16.12 provides a list of eight leading firms in the telecommunications equipment and services field. The new decade, 2000–2009, will be the decade of telecommunications and the internet. The amount of internet traffic is doubling every 100 days. The separate media of voice, video and data are converging onto a single unified network—the internet and corporate intranets. All this data and voice and video are converted to digital form and then transmitted via telecommunications highways. The information-carrying capacity of these networks is represented by the **bandwidth** which is the amount of information (bits) that can be transmitted in a period of time. Thus, the bandwidth (B) may be expressed in B bits per second where B is made as large as possible. The competition in the telecommunications industry is to manufacture equipment with very high bandwidth, B. Phone systems will transform to high-speed digital systems over the next decade and the eight companies we have selected will be among the leaders in that race. By 2004 there may be 800 million wireless phone users and 800 million internet users. Users will spend hours on the internet and the wireless phone, boosting the required capacity of communications lines. By 2003, there may be available a U.S. total long-distance bandwidth of 40 terabits where one terabit is 1,000 gigabits or roughly 15 million simultaneous phone calls. This bandwidth will be achieved by installing long-distance optical fiber systems and cable systems.

The system operators include AT&T, MCI-Worldcom, and SBC Communications. SBC is planning to merge with Ameritech in order to build nationwide service with about one-third of all the U.S. installed phone lines. MCI Worldcom has built a nationwide presence as one of the largest phone companies.

AT&T has agreed to merge with MediaOne in its race to build a telephone and cable network. This could transform AT&T into a broadband (large bandwidth) communications company.

Lucent Technologies builds equipment for telecommunication systems. Lucent is the equipment portion of the old AT&T, spinning off in 1996. Lucent has purchased Ascend in order to grow in the data communications equipment market. Lucent sales should grow annually at 13% or better over the next five years. This company has the power to generate annual returns of 30%.

Tellabs is an equipment maker and sells a leading digital connect system (DCS) which has emerged as the core element of the modern digital network. DCS are used in a variety of network situations where traffic management and monitoring are required. They commonly connect to the network side of a carrier's central office switch and provide a convenient and automated means to rearrange a carrier's network as traffic conditions and need change.

Qualcomm is the developer of the CDMA wireless technology and is a leader in the wireless industry. Code-division multiple access (CDMA) is a digital code which helps wireless phone providers to increase the number of subscribers in a given bandwidth. Qualcomm is competing with other wireless phone methods for standards acceptance. Qualcomm has great potential and may return 30% over the next five years.

Nokia is the world's largest manufacturer of cellular handsets. Nokia has rapidly gained share in the handset market, which is experiencing a growth rate of more than 40%. The company has released three new handset lines in 1997–1998, all of which have received enthusiastic acceptance by end users. The success of these products has put Nokia firmly in the lead in mobile phone sales. The company produced more than 40 million units in 1998.

JDS Uniphase is the result of a merger of JDS Fitel and Uniphase Corporation. JDS Fitel makes passive components for fiber-optic communications while Uniphase makes lasers and other active components used in fiber-optic networks. JDS Uniphase will be able to make value-added modules from their components, thus increasing their market value. We expect JDS Uniphase to grow revenues at a rate exceeding 30% per year.

## Utility Stocks

Table 16.13 lists six utility stocks that can help make up the core of a Tortoise portfolio. All six stocks are expected to return about 14% annually and provide significant defensive characteristics due to their yields. See Chapter 10 for discussion of these utility stocks.

## REIT Stocks

Table 16.14 lists three solid REIT stocks. All these REITs have a yield of at least 5% and should provide a return of 15% over a period of five to 10 years. See Chapter 10 for further discussion of REIT stocks.

## Contrarian Stocks

The basis of a Contrarian portfolio is two or three out of favor stocks. See Chapter 11 for a discussion of Contrarian stocks. We are looking for companies that have provided solid performance in the past but have experienced a down period due to a stumble in earnings or a drift away from former successes. Table 16.15 provides a list of four potential Contrarian stocks at the time of this writing.

Compaq Computer has stumbled on its internet sales channel and sales have leveled off. Furthermore, the company has experienced some trouble in absorbing the

**Table 16.13   Utilities Stocks (Projected 1999–2004)**

| Name | Industry | RoE (%) | G = GRE + Y (%) | Power | Value Line PCF | Value Index | Growth Rate of Dividends (%) | Expected Return (%) | Confidence Rank (%) | Probable Return (%) |
|---|---|---|---|---|---|---|---|---|---|---|
| Northwestern | Electric | 12 | 16.5 | 207 | 9.0 | 23.0 | 16.0 | 15 | 5 | 15 |
| Duke Energy | Electric & Gas | 14 | 13.4 | 244 | 9.0 | 27.1 | 2.0 | 14 | 5 | 14 |
| FPL Group | Electric | 13.5 | 14.0 | 227 | 8.0 | 28.4 | 4.0 | 14.0 | 5 | 14 |
| Nicor | Gas | 18 | 10.6 | 248 | 9 | 27.6 | 4.0 | 13.0 | 5 | 13 |
| American Water | Water | 12.5 | 12.6 | 180 | 9 | 20.0 | 7.0 | 13 | 5 | 13 |
| Reliant | Electric | 14.5 | 14 | 219 | 7.5 | 29.2 | 6.0 | 15 | 5 | 15 |

**Table 16.14   REIT Stocks (Projected 1999–2004)**

| Name | Industry | RoE (%) | G = GFFO + Y (%) | Power | Value Line PFFO | Value Index | Growth Rate of Dividends (%) | Expected Return (%) | Confidence Rank (%) | Probable Return (%) |
|---|---|---|---|---|---|---|---|---|---|---|
| Kimco | REIT | 11.5 | 16.0 | 205 | 12.0 | 17.1 | 9.0 | 15 | 5 | 15 |
| New Plan Excel | REIT | 12.5 | 18.0 | 270 | 10.0 | 27.0 | 10.0 | 15 | 5 | 15 |
| Weingarten | REIT | 15.5 | 13.0 | 222 | 10.5 | 21.1 | 10.0 | 15 | 4 | 15 |

**Table 16.15   Contrarian Stocks (Projected 1999–2004)**

| Name | Industry | Rcap (%) | G (%) | Power | Value Line PCF | Value Index | Growth Rate of Sales (%) | Expected Return (%) | Confidence Rank (%) |
|---|---|---|---|---|---|---|---|---|---|
| Compaq | Computer | 17.5 | 13 | 237 | 12.5 | 18.9 | 13.0 | 18 | 4 |
| Halliburton | Oil services | 11.0 | 13 | 103 | 14.0 | 7.4 | 10.0 | 15 | 4 |
| Motorola | Electric equipment | 12.0 | 9 | 97 | 10 | 9.7 | 7.0 | 15 | 4 |
| Disney | Leisure | 9.0 | 14.5 | 156 | 15 | 10.4 | 12.0 | 15 | 4 |

acquisition of Digital Equipment Corporation. Nevertheless, Compaq has a solid past record and should be able to return to its former glories.

Halliburton is an oil service company and experienced a decline due to low oil prices and decreased oil drilling.

Motorola is a turnaround company in the electronics and communications industry. Disney is a provider of movies and theme parks that has recently disappointed with modest earnings and revenue growth. All four companies may straighten out their troubles and provide a return of 15% or more over the next five years.

# 17 100 Millennium Stocks

In this chapter we provide the essential information for the 100 Millennium Stocks selected in Chapter 16. Contact Information is provided for each company and it is indicated if the company has a dividend reinvestment plan. Also a direct initial purchase plan is indicated for about 20 of the companies. This option allows the purchase of the initial shares directly from the company, thus bypassing a broker.

We also indicate the portfolio that each stock is suitable for. We identify 33 Gazelle stocks, 52 Hare stocks, 10 Tortoise stocks, and 5 Contra stocks.

Name: __Abbott Laboratories__    Symbol: __ABT__    Exchange: __NYSE__

Industry: __Health Care Products__          Stock Type: __Hare__

Profile: __Makes drugs, tests and medical instruments__

Address: __100 Abbott Park Rd., Abbott Park, IL 60064__

CEO: __D.L. Burnham and Miles D. White__    CFO: __Gary Coughlan__

Phone Number: __(847) 937-6100__    Investor Relations: __(847) 937-7300__

Internet Address: __www.abbott.com__

Dividend Reinvestment Plan: __Yes__    Contact: __(888) 332-2268__

Direct Purchase Plan: __No__

Average Annual Return '93-'98: __29.6 %__

---

Name: __Albertson's__    Symbol: __ABS__    Exchange: __NYSE__

Industry: __Retail-Food & Drug Stores__          Stock Type: __Hare__

Profile: __Operates retail grocery stores__

Address: __Box 20, Boise, ID 83726__

CEO: __Gary G. Michael__    CFO: __A. Craig Olson__

Phone Number: __(208) 395-6200__

Internet Address: __www.albertsons.com__

Dividend Reinvestment Plan: __Yes__    Contact: __(800) 982-7649__

Direct Purchase Plan: __No__

Average Annual Return '93-'98: __20.8 %__

Name: ___Alcoa___    Symbol: ___AA___    Exchange: ___NYSE___

Industry: ___Aluminum___    Stock Type: ___Hare___

Profile: ___World's largest producer of aluminum___

Address: ___201 Isabella St., Pittsburgh, PA 15212___

CEO: ___Paul O'Neill___    CFO: ___Richard Kelson___

Phone Number: ___(412) 553-3042___  or  ___(800) 522-6757___

Internet Address: ___www.alcoa.com___

Dividend Reinvestment Plan: ___Yes___    Contact: ___(800) 317-4445___

Direct Purchase Plan: ___No___

Average Annual Return '93-'98: ___18.8 %___

---

Name: ___Allied Signal___    Symbol: ___ALD___    Exchange: ___NYSE___

Industry: ___Aerospace Manufacturer___    Stock Type: ___Hare___

Profile: ___Leader in aerospace merging with Honeywell in building controls___

Address: ___101 Columbia Rd., Morristown, NJ 07962-2245___

CEO: ___Lawrence Bossidy___    CFO: ___Richard Wallman___

Phone Number: ___(973) 455-2000___  Investor Relations: ___(800) 707-4555___

Internet Address: ___www.alliedsignal.com___

Dividend Reinvestment Plan: ___Yes___    Contact: ___(800) 432-0140___

Direct Purchase Plan: ___No___

Average Annual Return '93-'98: ___19.4 %___

Name:   America Online     Symbol:   AOL     Exchange:   NYSE

Industry:   Internet Services              Stock Type:   Gazelle

Profile:   Leading online service provider

Address:   22000 AOL Way, Dulles, VA  20166

CEO:   R.W. Pitman              CFO:   J.M. Kelley

Phone Number:   (703) 448-8700

Internet Address:   www.aol.com/corp

Dividend Reinvestment Plan:   No

Direct Purchase Plan:   No

Average Annual Return '93-'98:   143.1 %

---

Name:   Applied Materials   Symbol:   AMAT   Exchange:   Nasdaq

Industry:   Semiconductor Equipment        Stock Type:   Hare

Profile:   Global leader in manufacturing equipment for semiconductors

Address:   3050 Bowers Ave., Santa Clara, CA  95054

CEO:   James Morgan              CFO:   Joseph Bronson

Phone Number:   (408) 727-5555

Internet Address:   www.appliedmaterials.com

Dividend Reinvestment Plan:   No

Direct Purchase Plan:   No

Average Annual Return '93-'98:   34.5 %

Name: __American International__     Symbol: __AIG__     Exchange: __NYSE__

Industry: __Insurance__               Stock Type: __Hare__

Profile: __Worldwide provider of diverse insurance products__

Address: __70 Pine St., New York, NY 10270__

CEO: __Maurice Greenberg__         CFO: __Howard Smith__

Phone Number: __(212) 770-7000__     Investor Relations: __(212) 770-6293__

Internet Address: __www.aig.com__

Dividend Reinvestment Plan: __Yes__       Contact: __(800) 446-2617__

Direct Purchase Plan: __No__

Average Annual Return '93-'98: __30.5 %__

---

Name: __American Water Works__     Symbol: __AWK__     Exchange: __NYSE__

Industry: __Water Utility__             Stock Type: __Tortoise__

Profile: __Largest investor-owned water utility in U.S.__

Address: __1025 Laurel Oak Rd., Box 1770, Voorhees, NJ 08043__

CEO: __J. James Barr__          CFO: __Joseph Hartnett__

Phone Number: __(609) 346-8200__

Internet Address: __www.amwater.com__

Dividend Reinvestment Plan: __Yes__       Contact: __(800) 736-3001__

Direct Purchase Plan: __No__

Average Annual Return '93-'98: __21.9 %__

Name: ___AT&T_____ Symbol: ___T____ Exchange: _NYSE___

Industry: ___Telecommunications_____ Stock Type: ___Hare_____

Profile: _____Provider of long distance and cable communications_____

Address: ___32 Ave. of the Americas, New York, NY  10013_____

CEO: ___C. Michael Armstrong_____ CFO: ___Daniel Somers_____

Phone Number: ___(212) 387-5400____ Investor Relations: ___(800) 348-8288___

Internet Address: ___www.att.com_____

Dividend Reinvestment Plan: ___Yes_____ Contact: ___(800) 348-8288____

Direct Purchase Plan: _____No_____

Average Annual Return '93-'98: ____19.1 %____

---

Name: __Automatic Data Processing___ Symbol: ___AUD____ Exchange: _NYSE__

Industry: ___Computer Services_____ Stock Type: ___Hare_____

Profile: _____Data processing of payroll, tax, order entry and records for clients_____

Address: ___One ADP Blvd., Roseland, NJ  07068_____

CEO: ___Arthur Weinbach_____ CFO: ___Richard Haviland_____

Phone Number: ___(937) 994-5000____

Internet Address: ___www.adp.com_____

Dividend Reinvestment Plan: ___No_____

Direct Purchase Plan: _____No_____

Average Annual Return '93-'98: ____24.9 %____

Name: ___Avon_____   Symbol: __AVP___   Exchange: __NYSE__

Industry: ___Cosmetics_____   Stock Type: ___Hare_____

Profile: ____Maker and direct seller of cosmetics, jewelry and gift items____

Address: ___1345 Ave. of the Americas, New York, NY 10105_____

CEO: ___Charles Perrin_____   CFO: ___Robert Corti_____

Phone Number: ___(212) 282-5000___   Investor Relations: ___(212) 282-5320___

Internet Address: ___www.avon.com_____

Dividend Reinvestment Plan: ___Yes_____   Contact: ___(201) 324-0498___

Direct Purchase Plan: _____No_____

Average Annual Return '93-'98: ___32.8 %___

---

Name: ___Bed, Bath & Beyond_____   Symbol: __BBBY___   Exchange: _Nasdaq_

Industry: ___Home Furnishings & Goods Retailer___   Stock Type: ___Gazelle____

Profile: ____Operates 190 stores selling domestic furnishings and goods_____

Address: ___650 Liberty Ave, Union, NJ 07083_____

CEO: ___Warren Eisenberg_____   CFO: ___Ronald Curwin_____

Phone Number: ___(908) 688-0888_____

Internet Address: ___www.bbby.com_____

Dividend Reinvestment Plan: ___No_____

Direct Purchase Plan: _____No_____

Average Annual Return '93-'98: ___31.7 %___

Name: __Boeing_____ Symbol: __BA___ Exchange: __NYSE__

Industry: __Aircraft Manufacturer_____ Stock Type: __Contrarian__

Profile: ____World's leading manufacturer of commercial jet aircraft_____

Address: __7755 East Marginal Way South, Seattle, WA 98108_____

CEO: __Philip Condit_____ CFO: __Deborah Hopkins__

Phone Number: __(206) 655-2121__ Investor Relations: __(800) 457-7723__

Internet Address: __www.boeing.com_____

Dividend Reinvestment Plan: __Yes_____ Contact: __(888) 777-0923__

Direct Purchase Plan: _____No_____

Average Annual Return '93-'98: __10.2 %__

---

Name: __Capital One Financial__ Symbol: __COF__ Exchange: __NYSE__

Industry: __Consumer Landing_____ Stock Type: __Gazelle__

Profile: ____Provides credit card loans_____

Address: __2980 Fairview Park Dr., Falls Church, VA 22042_____

CEO: __Richard Fairbank_____ CFO: __James Donehey__

Phone Number: __(703) 205-1000__ Investor Relations: __(703) 205-1039__

Internet Address: __www.capitalone.com_____

Dividend Reinvestment Plan: __Yes_____ Contact: __(800) 446-2617__

Direct Purchase Plan: _____No_____

Average Annual Return '93-'98: __70.2 %__ (3 years 1995-98)

Name: ___Carlisle___   Symbol: ___CSL___   Exchange: ___NYSE___

Industry: ___Plastics & Rubber Manufacturing___   Stock Type: ___Hare___

Profile: ___Diversified manufacturer of plastics, coatings and auto products___

Address: ___250 So. Clinton St., Syracuse, NY 13202___

CEO: ___Stephen Munn___   CFO: ___Robert Ryan___

Phone Number: ___(315) 474-2500___

Internet Address: ___www.carlisle.com___

Dividend Reinvestment Plan: ___Yes___   Contact: ___(800) 897-9071___

Direct Purchase Plan: ___No___

Average Annual Return '93-'98: ___27.6 %___

---

Name: ___Carnival___   Symbol: ___CCL___   Exchange: ___NYSE___

Industry: ___Leisure-Cruises___   Stock Type: ___Hare___

Profile: ___World's largest cruise provider___

Address: ___3655 N.W. 87th Ave., Miami, FL 33178___

CEO: ___Micky Arison___   CFO: ___Gerald Cahill___

Phone Number: ___(305) 599-2600___

Internet Address: ___www.carnivalcorp.com/investor___

Dividend Reinvestment Plan: ___Yes___   Contact: ___(800) 829-8432___

Direct Purchase Plan: ___No___

Average Annual Return '93-'98: ___33.9 %___

Name: ___Central Newspapers___   Symbol: __ECP__   Exchange: __NYSE__

Industry: ___Publications_____   Stock Type: ___Hare___

Profile: ___Publishes newspapers in Phoenix and Indianapolis___

Address: ___200 East Van Buren St., Phoenix, AZ  85004___

CEO: ___Louis Weil_____   CFO: ___Thomas Mac Gillivray___

Phone Number: ___(602) 444-1100___

Internet Address: ___www.centralnewspapers.com___

Dividend Reinvestment Plan: ___No___

Direct Purchase Plan: ___No___

Average Annual Return '93-'98: ___21.8 %___

---

Name: ___Champion Enterprises___   Symbol: __CHB__   Exchange: __NYSE__

Industry: ___Manufactured Buildings___   Stock Type: ___Hare___

Profile: ___Leading maker and retailer of manufactured housing___

Address: ___2701 University Drive, Auburn Hills, MI  48326___

CEO: ___Walter Young, Jr._____   CFO: ___Joseph Stegmayer___

Phone Number: ___(248) 340-9090___

Internet Address: ___www.champent.com___

Dividend Reinvestment Plan: ___No___

Direct Purchase Plan: ___No___

Average Annual Return '93-'98: ___44.1 %___

Name: ___Cisco Systems___  Symbol: __CSCO__  Exchange: __Nasdaq__

Industry: ___Computer Hardware___  Stock Type: ___Gazelle___

Profile: ___Provides hardware and software for network computing___

Address: ___170 West Tasman Dr., San Jose, CA 95134___

CEO: ___John Chambers___  CFO: ___Larry Carter___

Phone Number: ___(408) 526-4000___ or ___(800) 553-6387___

Internet Address: ___www.cisco.com___

Dividend Reinvestment Plan: ___No___

Direct Purchase Plan: ___No___

Average Annual Return '93-'98: ___66.8 %___

---

Name: ___Clorox___  Symbol: __CLX__  Exchange: __NYSE__

Industry: ___Home Care___  Stock Type: ___Hare___

Profile: ___Maker of household products___

Address: ___1221 Broadway, Oakland, CA 99612___

CEO: ___G. Craig Sullivan___  CFO: ___Karen Rose___

Phone Number: ___(510) 271-7000___  Investor Relations: ___(510) 271-2150___

Internet Address: ___www.clorox.com___

Dividend Reinvestment Plan: ___Yes___  Contact: ___(888) 259-6973___

Direct Purchase Plan: ___No___

Average Annual Return '93-'98: ___37.4 %___

Name:   Coca Cola _____         Symbol:   __KO__     Exchange:  _NYSE_

Industry:   __Beverages_____         Stock Type:  ___Hare___

Profile:   __World's largest soft drink company_____

Address:   __One Coca-Cola Plaza, Atlanta, GA  30313_____

CEO:  __M. Douglas Invester_____     CFO:   __James Chestnut___

Phone Number:  __(404) 676-2121___   Investor Relations:  __(404) 676-5766_

Internet Address:  ___www.thecoca-colacompany.com____

Dividend Reinvestment Plan:  __Yes_____   Contact:  __(888) 265-3747__

Direct Purchase Plan:  _____No_____

Average Annual Return '93-'98:  ____26.1 %____

---

Name:  __Compaq Computer_____         Symbol:  _____   Exchange:  _NYSE_

Industry:  __Computer Hardware_____         Stock Type:  __Contrarian__

Profile:   __Produces PC computers and servers_____

Address:   __20555 State Highway 249, Houston, TX  77070_____

CEO:  __Benjamin Rosen_____     CFO:  __N.A.___

Phone Number:  __(281) 370-0670___

Internet Address:  __www.compaq.com_____

Dividend Reinvestment Plan:  __Yes_____   Contact:  __(888) 218-4373_

Direct Purchase Plan:  _____Yes_____   Contact:  __(888) 218-4373_

Average Annual Return '93-'98:  ____53.6 %____

Name: __Computer Associates__    Symbol: __CA__    Exchange: __NYSE__

Industry: __Computer Software__      Stock Type: __Hare__

Profile: __Leading independent software company__

Address: __One Computer Associates Plaza, Islandia, NY 11788__

CEO: __Charles B. Wang__      CFO: __Peter Schwartz__

Phone Number: __(516) 342-5224__

Internet Address: __www.cai.com__

Dividend Reinvestment Plan: __Yes__      Contact: __(800) 244-7155__

Direct Purchase Plan: __No__

Average Annual Return '93-'98: __29.5 %__

---

Name: __Dell Computer__    Symbol: __DELL__    Exchange: __Nasdaq__

Industry: __Computer Hardware__      Stock Type: __Gazelle__

Profile: __Makes and sells direct PC computers and servers__

Address: __One Dell Way, Round Rock, TX 78682__

CEO: __Michael Dell__      CFO: __Thomas Meredith__

Phone Number: __(512) 338-4400__    Investor Relations: __(512) 728-8315__

Internet Address: __www.dell.com__

Dividend Reinvestment Plan: __No__

Direct Purchase Plan: __No__

Average Annual Return '93-'98: __152.9 %__

Name: ___Disney_____ Symbol: _DIS__ Exchange: _New York_

Industry: ___Entertainment_____ Stock Type: ___Contrarian__

Profile: ___Operates theme parks and makes movies and associated products___

Address: ___Box 11447, Burbank, CA  91510_____

CEO: ___Michael Eisner_____ CFO: ___Thomas Skaggs___

Phone Number: ___(818) 553-7200___ Investor Relations: ___(818) 553-7200__

Internet Address: ___www.disney.com_____

Dividend Reinvestment Plan: ___Yes_____ Contact: ___(800) 948-2222_

Direct Purchase Plan: _____Yes_____ Contact: ___(800) 948-2222_

Average Annual Return '93-'98: ___16.9 %___

---

Name: ___Dollar General_____ Symbol: _DG____ Exchange: _NYSE_

Industry: ___Discount Stores_____ Stock Type: ___Hare____

Profile: ___Operates 3,700 discount stores_____

Address: ___104 Woodmont Blvd., Nashville, TN  37205_____

CEO: ___Cal Turner_____ CFO: ___Phil Richards_____

Phone Number: ___(615) 783-2000___

Internet Address: ___www.dollargeneral.com_____

Dividend Reinvestment Plan: ___No_____

Direct Purchase Plan: _____No_____

Average Annual Return '93-'98: ___31.1 %___

Name: __Duke Energy__     Symbol: __DUK__     Exchange: __NYSE__

Industry: __Electric & Gas Utility__          Stock Type: __Tortoise__

Profile: __Electric and gas utility in No. and So. Carolina__

Address: __526 So. Church St., Charlotte, NC 28202__

CEO: __Richard Priory__          CFO: __Richard Osborne__

Phone Number: __(704) 594-6200__

Internet Address: __www.duke-energy.com__

Dividend Reinvestment Plan: __Yes__          Contact: __(800) 488-3853__

Direct Purchase Plan: __Yes__

Average Annual Return '93-'98: __13.6 %__

---

Name: __Ecolab__     Symbol: __ECL__     Exchange: __NYSE__

Industry: __Cleaning Products__          Stock Type: __Hare__

Profile: __Provides cleaning and maintenance products__

Address: __370 Wabasha St., N., St. Paul, MN 55102__

CEO: __Allan Schuman__          CFO: __Michael Shannon__

Phone Number: __(651) 293-2233__

Internet Address: __www.ecolab.com__

Dividend Reinvestment Plan: __Yes__          Contact: __(800) 322-8325__

Direct Purchase Plan: __No__

Average Annual Return '93-'98: __28.5 %__

Name: ___El Paso Energy_____          Symbol: ___EPG___   Exchange: ___NYSE___

Industry: ___Natural Gas_____          Stock Type: ___Tortoise___

Profile: ___Leading natural gas company merging with Sonat_____

Address: ___1001 Louisiana St., Houston, TX 77002_____

CEO: ___William Wise_____   CFO: ___H. Brent Austin_____

Phone Number: ___(713) 420-2131___   Investor Relations: ___(888) 202-9971___

Internet Address: ___www.epenergy.com_____

Dividend Reinvestment Plan: ___Yes_____   Contact: ___(800) 736-3001___

Direct Purchase Plan: _____No_____

Average Annual Return '93-'98: ___17.9 %___


Name: ___EMC_____          Symbol: ___EMC___   Exchange: ___NYSE___

Industry: ___Computer Storage_____          Stock Type: ___Gazelle___

Profile: ___Makes high performance, large storage products_____

Address: ___35 Parkwood Dr., Hopkinton, MA 01748_____

CEO: ___Michael Ruettgers_____   CFO: ___William Teuber_____

Phone Number: ___(508) 435-1000___

Internet Address: ___www.emc.com_____

Dividend Reinvestment Plan: ___No_____

Direct Purchase Plan: _____No_____

Average Annual Return '93-'98: ___59.4 %___

Name: __Equifax__ Symbol: __EFX__ Exchange: _____

Industry: __Financial Information__ Stock Type: __Hare__

Profile: __Provides credit reports and other financial information services__

Address: __1900 Peachtree St., NW, Atlanta, GA 30309__

CEO: __Thomas Chapman__ CFO: __David Post__

Phone Number: __(404) 885-8000__

Internet Address: __www.equifax.com__

Dividend Reinvestment Plan: __Yes__ Contact: __(800) 568-3476__

Direct Purchase Plan: __Yes__ Contact: __(800) 887-2971__

Average Annual Return '93-'98: __24.6 %__

Name: __Ethan Allen__ Symbol: __ETH__ Exchange: __NYSE__

Industry: __Furniture Manufacture and sales__ Stock Type: __Hare__

Profile: __Maker and Retailer of home furniture__

Address: __Ethan Allen Dr., Box 1966, Danbury, CT 06811__

CEO: __M. Farooq Kathwari__ CFO: __Gevado Burdo__

Phone Number: __(203) 743-8000__

Internet Address: __www.ethanallen.com__

Dividend Reinvestment Plan: __No__

Direct Purchase Plan: __No__

Average Annual Return '93-'98: __27.2 %__

Name: ___Exxon___          Symbol: ___XON___     Exchange: ___NYSE___

Industry: ___Petroleum___                    Stock Type: ___Hare___

Profile: ___Leading integrated oil company merging with Mobil___

Address: ___5959 Las Colinas Blvd., Irving, TX  75039-2298___

CEO: ___Lee R. Raymond___          CFO: ___F. Risch___

Phone Number: ___(972) 444-1000___

Internet Address: ___www.exxon.com___

Dividend Reinvestment Plan: ___Yes___     Contact: ___(800) 252-1800___

Direct Purchase Plan: ___Yes___

Average Annual Return '93-'98: ___22.6 %___

---

Name: ___Fannie Mae___          Symbol: ___FNM___     Exchange: ___NYSE___

Industry: ___Mortgage financing___                    Stock Type: ___Hare___

Profile: ___Nation's largest provider of residential mortgage funds___

Address: ___3900 Wisconsin Ave., NW, Washington, DC  20016-2892___

CEO: ___Franklin Raines___          CFO: ___J. Timothy Howard___

Phone Number: ___(202) 752-7115___

Internet Address: ___www.fanniemae.com___

Dividend Reinvestment Plan: ___Yes___     Contact: ___(888) 910-8277___

Direct Purchase Plan: ___Yes___     Contact: ___(888) 289-3266___

Average Annual Return '93-'98: ___33.6 %___

Name: __FDX__ Symbol: __FDX__ Exchange: __NYSE__

Industry: __Airfreight Courier__ Stock Type: __Hare__

Profile: __Provides overnight door to door delivery services__

Address: __6075 Poplar Ave., Memphis, TN 38119__

CEO: __Frederick Smith__ CFO: __Alan Graf, Jr.__

Phone Number: __(901) 369-3600__

Internet Address: __www.fedex.com__

Dividend Reinvestment Plan: __No__

Direct Purchase Plan: __No__

Average Annual Return '93-'98: __20.3 %__

---

Name: __Firstar__ Symbol: __FSR__ Exchange: __NYSE__

Industry: __Bank__ Stock Type: __Gazelle__

Profile: __Banking and financial services__

Address: __777 E. Wisconsin Ave., Milwaukee, WI 53202__

CEO: __Jerry Grundhofer__ CFO: __David Moffett__

Phone Number: __(414) 765-5235__

Internet Address: __www.firstar.com__

Dividend Reinvestment Plan: __Yes__ Contact: __(800) 637-7549__

Direct Purchase Plan: __No__

Average Annual Return '93-'98: __55.3 %__

Name: ___First Union_____ Symbol: ___FTU___ Exchange: ___NYSE___

Industry: ___Bank_____ Stock Type: ___Hare_____

Profile: _____Large regional bank in eastern U.S._____

Address: ___One First Union Center, Charlotte, NC 28288-0570_____

CEO: ___Edward Crutchfield_____ CFO: ___Robert Atwood_____

Phone Number: ___(704) 374-2137_____

Internet Address: ___www.firstunion.com_____

Dividend Reinvestment Plan: ___Yes_____ Contact: ___(800) 347-1246_____

Direct Purchase Plan: _____No_____

Average Annual Return '93-'98: _____28.4 %_____

---

Name: ___FPL Group_____ Symbol: ___FPL___ Exchange: ___NYSE___

Industry: ___Electric Utility_____ Stock Type: ___Tortoise___

Profile: _____Electric utility in Florida_____

Address: ___700 Universe Blvd., Juno Beach, FL 33408-0420_____

CEO: ___James Broadhead_____ CFO: ___Mary Lou Kromer_____

Phone Number: ___(800) 222-4511_____

Internet Address: ___www.fplgroup.com_____

Dividend Reinvestment Plan: ___Yes_____ Contact: ___(888) 218-4392_____

Direct Purchase Plan: _____No_____

Average Annual Return '93-'98: _____14.4 %_____

Name: ___GAP_____   Symbol: __GPS___   Exchange: __NYSE__

Industry: ___Apparel Stores_____   Stock Type: ___Gazelle___

Profile: ____Operates 2,600 casual apparel stores_____

Address: ___One Hamson St., San Francisco, CA 94105_____

CEO: ___Millard Drexler_____   CFO: ___W. Hashagen_____

Phone Number: __(415) 427-2000____

Internet Address: ___www.gap.com_____

Dividend Reinvestment Plan: ___No_____

Direct Purchase Plan: _____No_____

Average Annual Return '93-'98: ____46.3 %____

---

Name: ___General Electric_____   Symbol: __GE___   Exchange: __NYSE__

Industry: ___Electrical Equipment_____   Stock Type: __Hare_____

Profile: ____Leading diversified manufacturing company_____

Address: ___3135 Easton Turnpike Fairfield, CT 06431_____

CEO: ___John Welch_____   CFO: ___Keith Sherin_____

Phone Number: ___(203) 373-2211____

Internet Address: ___www.ge.com_____

Dividend Reinvestment Plan: ___Yes_____   Contact: ___(800) 786-2543___

Direct Purchase Plan: _____Yes_____

Average Annual Return '93-'98: ____34.2 %____

Name: ___Guidant_____        Symbol: __GDT___   Exchange: _NYSE__

Industry: ___Medical Equipment_____        Stock Type: ___Gazelle___

Profile: ____Makes medical equipment_____

Address: ___111 Monument Circle, Indianapolis, IN 46204-5129_____

CEO: ___Ronald Dollens_____   CFO: ___Keith Brauer_____

Phone Number: ___(317) 971-2000_____

Internet Address: ___www.guidant.com_____

Dividend Reinvestment Plan: ___Yes_____   Contact: ___(800) 317-4445___

Direct Purchase Plan: _____Yes_____   Contact: ___(800) 537-1677___

Average Annual Return '93-'98: ____73.5 %____   (3 years 1995-98)

---

Name: ___Halliburton_____        Symbol: __HAL___   Exchange: _NYSE__

Industry: ___Energy Services_____        Stock Type: ___Contrarian__

Profile: ____Provides oil drilling and pumping services_____

Address: ___3600 Lincoln Plaza, 500 N. Akard St., Dallas, TX 75201____

CEO: ___Richard Cheney_____   CFO: ___Gary Morris_____

Phone Number: ___(214) 978-2600____   or   ____(888) 669-3920_____

Internet Address: ___www.halliburton.com_____

Dividend Reinvestment Plan: ___Yes_____   Contact: ___(800) 279-1227___

Direct Purchase Plan: _____No_____

Average Annual Return '93-'98: ____16.6 %____

Name:　　Harley Davidson　　　　　Symbol:　HDI　　Exchange:　NYSE

Industry:　　Motorcycles　　　　　　　　　Stock Type:　　Hare

Profile:　　　Maker of heavyweight motorcycles

Address:　　3700 W. Juneau Ave., P.O. Box 653, Milwaukee, WI 53201

CEO:　　Jeffrey Bleustein　　　　　　CFO:　　James Ziemer

Phone Number:　(414) 342-4680　　　or　　(877) 437-8625

Internet Address:　　www.harley-davidson.com

Dividend Reinvestment Plan:　Yes　　　　Contact:　(800) 637-7549

Direct Purchase Plan:　　　No

Average Annual Return '93-'98:　　34.6 %

---

Name:　　Herman Miller　　　　　Symbol:　MLHR　　Exchange:　Nasdaq

Industry:　　Office Furniture　　　　　　　Stock Type:　　Hare

Profile:　　Designs, makes, and sells office furniture

Address:　　Box 302, Zeeland, MI 49464

CEO:　　David Nelson　　　　　　　CFO:　　Brian Walker

Phone Number:　(616) 654-3000

Internet Address:　　www.hermanmiller.com

Dividend Reinvestment Plan:　Yes　　　　Contact:　(800) 446-2617

Direct Purchase Plan:　　　No

Average Annual Return '93-'98:　　30.3 %

Name: ___Home Depot___          Symbol: ___HD___     Exchange: ___NYSE___

Industry: ___Hardware & Appliances Stores___     Stock Type: ___Gazelle___

Profile: ___Operator of chain of retail building supply stores___

Address: ___2455 Paces Ferry Rd., NW, Atlanta, GA 30339___

CEO: ___Arthur Blank___               CFO: ___Dennis Carey___

Phone Number: ___(770) 433-8211___

Internet Address: ___www.homedepot.com___

Dividend Reinvestment Plan: ___Yes___     Contact: ___(800) 577-0177___

Direct Purchase Plan: ___Yes___          Contact: ___(800) 928-0380___

Average Annual Return '93-'98: ___36.5 %___

---

Name: ___Illinois Tool Works___     Symbol: ___ITW___     Exchange: ___NYSE___

Industry: ___Metal Tools and Parts___          Stock Type: ___Hare___

Profile: ___Makes metal tools, components and fasteners___

Address: ___3600 w. Lake Ave., Glenview, IL 60025___

CEO: ___W. James Farrell___            CFO: ___Jon Kinney___

Phone Number: ___(847) 724-7500___

Internet Address: ___itwinc.com___

Dividend Reinvestment Plan: ___Yes___     Contact: ___(888) 829-7424___

Direct Purchase Plan: ___No___

Average Annual Return '93-'98: ___25.8 %___

Name: ___Intel_____ Symbol: ___INTC___ Exchange: __Nasdaq__

Industry: ___Semiconductors_____ Stock Type: ___Gazelle____

Profile: ____Designer and manufacturer of integrated circuits_____

Address: ___Box 58119, Santa Clara, CA 95052_____

CEO: ___Craig Barrett_____ CFO: ___Andy Bryant_____

Phone Number: ____(408) 765-8080____

Internet Address: ___www.intel.com_____

Dividend Reinvestment Plan: ___Yes_____ Contact: ___(800) 298-0146___

Direct Purchase Plan: _____No_____

Average Annual Return '93-'98: _____50.6 %____

---

Name: __International Business Machines__ Symbol: __IBM__ Exchange: __NYSE__

Industry: ___Computers_____ Stock Type: ___Hare_____

Profile: ____World's leader of information system equipment and services_____

Address: ___New Orchard Rd., Armonk, NY 10504_____

CEO: ___Louis Gerstner, Jr._____ CFO: ___Douglas Maine_____

Phone Number: ____(914) 499-1900_____ or ____(800) 426-4968_____

Internet Address: ___www.ibm.com_____

Dividend Reinvestment Plan: ___Yes_____ Contact: ___(888) 421-8860___

Direct Purchase Plan: _____Yes_____ Contact: ___(888) 421-8860___

Average Annual Return '93-'98: _____47.1 %____

Name: ___Interpublic Group___     Symbol: __IPG__     Exchange: __NYSE__

Industry: ___Advertising___                    Stock Type: ___Hare___

Profile: ___Worldwide organization of advertising and marketing companies___

Address: ___1271 Ave. of the Americas, New York, NY 10020___

CEO: ___Philip Geier, Jr.___          CFO: ___Eugene Beard___

Phone Number: ___(212) 399-8000___

Internet Address: ___www.interpublic.com___

Dividend Reinvestment Plan: ___Yes___     Contact: ___(201) 324-0498___

Direct Purchase Plan: ___No___

Average Annual Return '93-'98: ___32.0 %___

---

Name: ___JDS Uniphase___     Symbol: __JDSU__     Exchange: __Nasdaq__

Industry: ___Telecommunications Electronics___     Stock Type: ___Gazelle___

Profile: ___Manufacturer of optoelectronics devices for telecomm. systems___

Address: ___163 Baypointe Pway, San Jose, CA 95134___

CEO: ___Kevin Kalkhoven___          CFO: ___Anthony Muller___

Phone Number: ___(408) 434-1800___

Internet Address: ___www.uniphase.com___

Dividend Reinvestment Plan: ___No___

Direct Purchase Plan: ___No___

Average Annual Return '93-'98: ___114 %___

Name:  Johnson and Johnson     Symbol:  JNJ     Exchange:  NYSE

Industry:  Health Care Products     Stock Type:  Hare

Profile:  Makes consumer and profession health care products

Address:  One Johnson & Johnson Plaza, New Brunswick, NJ 08933

CEO:  Ralph Larsen     CFO:  Robert Darretta

Phone Number:  (732) 524-0400     or     (800) 950-5089

Internet Address:  www.jnj.com

Dividend Reinvestment Plan:  Yes     Contact:  (800) 328-9033

Direct Purchase Plan:  No

Average Annual Return '93-'98:  32.4 %

---

Name:  Johnson Controls     Symbol:  JCI     Exchange:  NYSE

Industry:  Automotive Parts & Controls     Stock Type:  Hare

Profile:  Maker of auto parts and control devices

Address:  Box 591, Milwaukee, WI 53201

CEO:  James Keyes     CFO:  Stephen Roell

Phone Number:  (414) 228-1200

Internet Address:  www.johnsoncontrols.com

Dividend Reinvestment Plan:  Yes     Contact:  (800) 828-1489

Direct Purchase Plan:  Yes     Contact:  (800) 524-6220

Average Annual Return '93-'98:  20.0 %

Name: ___Jones Apparel___    Symbol: ___JNY___    Exchange: ___NYSE___

Industry: ___Apparel___    Stock Type: ___Hare___

Profile: ___Designer and Marketer of women's clothing___

Address: ___250 Rittenhouse Circle, Bristol, PA  19007___

CEO: ___Sidney Kimmel___    CFO: ___Wesley Card___

Phone Number: ___(215) 785-4000___

Internet Address: ___www.jny.com___

Dividend Reinvestment Plan: ___No___

Direct Purchase Plan: ___No___

Average Annual Return '93-'98: ___24.2 %___

---

Name: ___Kimberly Clark___    Symbol: ___KMB___    Exchange: ___NYSE___

Industry: ___Household Products-paper___    Stock Type: ___Hare___

Profile: ___Maker of personal care products___

Address: ___Box 619100, Dallas, TX  75261-9100___

CEO: ___Wayne Sanders___    CFO: ___John Donehower___

Phone Number: ___(972) 281-1200___   or   ___(800) 639-1352___

Internet Address: ___www.kimberly-Clark.com___

Dividend Reinvestment Plan: ___Yes___    Contact: ___(800) 730-4001___

Direct Purchase Plan: ___No___

Average Annual Return '93-'98: ___19.6 %___

Name: ___Kimco Realty_____ Symbol: __KIM___ Exchange: __NYSE__

Industry: ___REIT_____ Stock Type: ___Tortoise____

Profile: ____REIT holding shopping center properties_____

Address: ___3333 New Hyde Park Dr., New Hyde Park, NY 11042_____

CEO: ___Milton Cooper_____ CFO: ___Michael Pappagallo____

Phone Number: ___(516) 869-9000___

Internet Address: ___www.kimcorealty.com_____

Dividend Reinvestment Plan: ___Yes_____ Contact: ___(781) 575-3400___

Direct Purchase Plan: _____No_____

Average Annual Return '93-'98: ____17.7 %____

---

Name: ___Linear Technology_____ Symbol: __LLTC___ Exchange: __Nasdaq__

Industry: ___Integrated Circuits_____ Stock Type: ___Gazelle____

Profile: ____Makes high performance linear integrated circuits_____

Address: ___1630 McCarthy Blvd., Milpitas, CA 95035_____

CEO: ___Robert Swanson_____ CFO: ___Paul Coghlan_____

Phone Number: ___(408) 432-1900___

Internet Address: ___www.linear-tech.com_____

Dividend Reinvestment Plan: ___No_____

Direct Purchase Plan: _____No_____

Average Annual Return '93-'98: ____36.5 %____

Name: ___Lucent_____     Symbol: ___LU___     Exchange: __NYSE__

Industry: ___Telecommunications Equipment____     Stock Type: ___Gazelle____

Profile: ____Leading maker of telecom equipment_____

Address: ___600 Mountain Ave, NJ  07974_____

CEO: ___Richard McGuinn_____     CFO: ___Donald Peterson_____

Phone Number: ___(888) 458-2368____

Internet Address: ___www.lucent.com_____

Dividend Reinvestment Plan: ___Yes_____     Contact: ___(888) 582-3686____

Direct Purchase Plan: _____Yes_____     Contact: ___(800) 774-4117____

Average Annual Return '93-'98: ___55.0 %____     (2 years '96-98)


Name: ___Marriott International___     Symbol: __MAR__     Exchange: __NYSE__

Industry: ___Hotels_____     Stock Type: ___Hare___

Profile: ____Operator of hotels and resorts_____

Address: ___Marriott Drive, Washington, DC  20058_____

CEO: ___J.W. Marriott, Jr._____     CFO: ___Arne Sorenson_____

Phone Number: ___(301) 380-6500____

Internet Address: ___www.marriott.com_____

Dividend Reinvestment Plan: ___Yes_____     Contact: ___(800) 519-3111____

Direct Purchase Plan: _____No_____

Average Annual Return '93-'98: ___NA___

Name: ___Maytag_____     Symbol: _MYG__   Exchange: _NYSE__

Industry: ___Electrical Appliances_____     Stock Type: ___Hare____

Profile: ___Makes home and commercial appliances_____

Address: ___403 W. Fourth St., No., Newton, IA  50208_____

CEO: ___Leonard Hadley_____     CFO: ___Gerald Pribanic____

Phone Number: ___(515) 792-7000___

Internet Address: ___www.maytag.com_____

Dividend Reinvestment Plan: ___Yes_____   Contact: ___(888) 237-0935___

Direct Purchase Plan: _____No_____

Average Annual Return '93-'98: ___31.4 %_____

---

Name: ___MBNA_____     Symbol: ___KRB___   Exchange: _NYSE__

Industry: ___Consumer Lender_____     Stock Type: ___Gazelle___

Profile: ___World's largest independent credit card lender_____

Address: ___Wilmington, DE  19884-0131_____

CEO: ___Alfred Lerner_____     CFO: ___Scott Kaufman____

Phone Number: ___(800) 362-6255_____

Internet Address: ___www.mbnainternational.com_____

Dividend Reinvestment Plan: ___No_____

Direct Purchase Plan: _____No_____

Average Annual Return '93-'98: ___44.2 %_____

Name: ___MCI Worldcom___ Symbol: _WCOM_ Exchange: _Nasdaq_

Industry: ___Telecommunication Services___ Stock Type: ___Gazelle___

Profile: ___Provider of local, long distance and Internet telecom services___

Address: ___515 E. Amite St., Jackson, MS 39201___

CEO: ___Bernald Ebbers___ CFO: ___Scott Sullivan___

Phone Number: ___(877) 624-9266___

Internet Address: ___www.wcom.com___

Dividend Reinvestment Plan: ___No___

Direct Purchase Plan: ___No___

Average Annual Return '93-'98: ___42.9 %___

---

Name: ___Medtronic___ Symbol: ___MDT___ Exchange: ___NYSE___

Industry: ___Medical Equipment___ Stock Type: ___Gazelle___

Profile: ___World's largest manufacturer of implantable biomedical devices___

Address: ___7000 Central Ave., NE, Minneapolis, MN 55432___

CEO: ___William George___ CFO: ___Robert Ryan___

Phone Number: ___(612) 514-4000___

Internet Address: ___www.medtronic.com___

Dividend Reinvestment Plan: ___Yes___ Contact: ___(800) 468-9716___

Direct Purchase Plan: ___No___

Average Annual Return '93-'98: ___49.5 %___

Name: ___Mellon Bank___ Symbol: __MEL__ Exchange: __NYSE__

Industry: __Bank__ Stock Type: __Hare__

Profile: ___Banking and financial services___

Address: ___One Mellon Bank Center, Pittsburgh, PA 15258___

CEO: __Martin McGuinn__ CFO: ___Steven Elliott___

Phone Number: ___(412) 234-5601___ or ___(800) 205-7699___

Internet Address: ___www.mellon.com___

Dividend Reinvestment Plan: ___Yes___ Contact: ___(800) 842-7629___

Direct Purchase Plan: ___Yes___ Contact: ___(800) 205-7699___

Average Annual Return '93-'98: ___36.2 %___

---

Name: ___Merck___ Symbol: __MRK__ Exchange: __NYSE__

Industry: ___Pharmaceuticals___ Stock Type: __Hare__

Profile: ___Leading maker of pharmaceuticals___

Address: ___One Merck Drive Box 100, Whitehouse Station, NJ 08889___

CEO: __Raymond Gilmartin__ CFO: ___Judy Lewent___

Phone Number: ___(908) 423-1000___

Internet Address: ___www.merck.com___

Dividend Reinvestment Plan: ___Yes___ Contact: ___(800) 613-2104___

Direct Purchase Plan: ___Yes___ Contact: ___(800) 774-4117___

Average Annual Return '93-'98: ___37.0 %___

Name: ___Microsoft_____     Symbol: __MSFT___   Exchange: __Nasdaq__

Industry: ___Computer Software_____     Stock Type: ___Gazelle___

Profile: ____Leading maker of operating systems and PC applications_____

Address: ___One Microscoft Way, Redmond, WA  98052_____

CEO: ___Bill Gates_____     CFO: ___Gregory Maffei_____

Phone Number: ___(425) 936-4400_____   or   ____(800) 285-7772_____

Internet Address: ___www.microsoft.com_____

Dividend Reinvestment Plan: ___No_____

Direct Purchase Plan: _____No_____

Average Annual Return '93-'98: ____68.9 %____

---

Name: ___Motorola_____     Symbol: __MOT___   Exchange: __NYSE__

Industry: ___Communications Equipment_____     Stock Type: ___Contrarian__

Profile: ____Maker of electronic equipment and telecom equipment_____

Address: ___1303 E. Algonquin Rd., Schaumburg, IL  60196_____

CEO: ___C.B. Galvin_____     CFO: ___Carl Koenemann_____

Phone Number: ___(847) 576-5000___

Internet Address: ___www.motorola.com_____

Dividend Reinvestment Plan: ___Yes_____   Contact: ___(800) 704-4098_____

Direct Purchase Plan: _____No_____

Average Annual Return '93-'98: ____6.6 %____

Name:  __New Plan Excel Realty__    Symbol:  __NXL__    Exchange:  __NYSE__

Industry:  __REIT__                            Stock Type:  __Tortoise__

Profile:  __REIT holding shopping malls and apartments__

Address:  __1120 Ave. of the Americas, New York, NY  10036__

CEO:  __Arnold Laubich__              CFO:  __Dean Bernstein__

Phone Number:  __(212) 869-3000__

Internet Address:  __www.newplanexcel.com__

Dividend Reinvestment Plan:  __Yes__      Contact:  __(800) 730-6001__

Direct Purchase Plan:  __No__

Average Annual Return '93-'98:  __16.3 %__

---

Name:  __Nicor__                  Symbol:  __GAS__    Exchange:  __NYSE__

Industry:  __Natural Gas Utility__              Stock Type:  __Tortoise__

Profile:  __Natural gas utility in Illinois__

Address:  __Box 3014, Naperville, IL  60566__

CEO:  __Thomas Fisher__              CFO:  __David Cyrenoski__

Phone Number:  __(630) 305-9500__

Internet Address:  __www.nicorine.com__

Dividend Reinvestment Plan:  __Yes__      Contact:  __(630) 305-9500__

Direct Purchase Plan:  __No__

Average Annual Return '93-'98:  __13.3 %__

Name: ___Nokia_____     Symbol: __NOKA__     Exchange: __NYSE__

Industry: ___Telecomm. Equipment_____     Stock Type: ___Gazelle____

Profile: _____World's second largest supplies of mobile phones and networks_____

Address: ____Hdqtrs: Finland; U.S.: 6000 Connection Dr., Irving, TX 75039___

CEO: ___Jorma Ollila_____     U.S. Vice President: ___Martin Sandelin_____

Phone Number: ___(972) 894-4880___

Internet Address: ___www.nokia.com_____

Dividend Reinvestment Plan: ___No_____

Direct Purchase Plan: _____Yes_____     Contact: ___(800) 483-9010___

Average Annual Return '93-'98: ____81.3 %____

---

Name: ___Northwestern_____     Symbol: __NOR__     Exchange: __NYSE__

Industry: ___Electric & Natural Gas Utility_____     Stock Type: ___Tortoise___

Profile: _____Electric and gas utility with network communication services_____

Address: ___125 S. Dakota Ave., Sioux Falls, SD 57104___

CEO: ___Merle Lewis_____     CFO: ___Daniel Newell___

Phone Number: ___(605) 978-2908___

Internet Address: ___www.northwestern.com_____

Dividend Reinvestment Plan: ___Yes_____     Contact: ___(800) 677-6716___

Direct Purchase Plan: _____Yes_____

Average Annual Return '93-'98: ____19.0 %____

Name: ___Oracle_____ Symbol: __ORCL___ Exchange: __Nasdaq__

Industry: ___Computer Software_____ Stock Type: ___Gazelle____

Profile: ____Maker of database and application software_____

Address: ___500 Oracle Pway, Redwood City, CA 94065_____

CEO: ___Lawrence Ellison_____ CFO: ____Jeffrey Henley_____

Phone Number: ___(650) 506-7000___

Internet Address: ___www.oracle.com_____

Dividend Reinvestment Plan: ___No_____

Direct Purchase Plan: _____No_____

Average Annual Return '93-'98: ___38.3 %____

---

Name: ___Paychex_____ Symbol: __PAYX___ Exchange: __Nasdaq__

Industry: ___Business Services-Payroll_____ Stock Type: ___Gazelle____

Profile: ____Provides payroll and accounting services for clients_____

Address: ___911 Panorama Trail So., Rochester, NY 14625_____

CEO: ___Thomas Golisano_____ CFO: ____John Morphy_____

Phone Number: ___(716) 385-6666___

Internet Address: ___www.paychex.com_____

Dividend Reinvestment Plan: ___Yes_____ Contact: ___(800) 937-5449____

Direct Purchase Plan: _____No_____

Average Annual Return '93-'98: ___50.5 %____

Name:  ___Pfizer_____        Symbol: __PFE___    Exchange: __NYSE__

Industry: ___Pharmaceuticals_____        Stock Type: ___Gazelle____

Profile: ____Leading maker of pharmaceuticals_____

Address: ___235 E. 42nd St., New York, NY  10017_____

CEO: ___William Steere_____        CFO: ___David Shedparz_____

Phone Number: ___(212) 573-2323____

Internet Address: ___www.pfizer.com_____

Dividend Reinvestment Plan: ___Yes_____        Contact: ___(800) 733-9393____

Direct Purchase Plan: _____Yes_____

Average Annual Return '93-'98: ____51.2 %____

---

Name: ___Pitney Bowes_____        Symbol: __PBI___    Exchange: __NYSE__

Industry: ___Business Document Messaging_____        Stock Type: ___Hare_____

Profile: _____World's largest maker of postage meters and mailing equipment__

Address: ____1 Elecroft Rd., Stamford, CT  06926_____

CEO: ___Michael Critelli_____        CFO: ___Murray Reichenstein__

Phone Number: ___(203) 356-5000____

Internet Address: ___www.pitneybowes.com_____

Dividend Reinvestment Plan: ___Yes_____        Contact: ___(800) 648-8170____

Direct Purchase Plan: _____No_____

Average Annual Return '93-'98: ____29.3 %____

Name: ___Price, T. Rowe___ Symbol: ___TROW___ Exchange: ___Nasdaq___

Industry: ___Mutual Funds___ Stock Type: ___Hare___

Profile: ___Investment advice and mutual funds___

Address: ___100 E. Pratt St., Baltimore, MD 21202___

CEO: ___George Roche___ CFO: ___Alvin Younger___

Phone Number: ___(410) 345-2000___

Internet Address: ___www.troweprice.com___

Dividend Reinvestment Plan: ___No___

Direct Purchase Plan: ___No___

Average Annual Return '93-'98: ___38.3 %___

---

Name: ___Providian___ Symbol: ___PVN___ Exchange: ___NYSE___

Industry: ___Consumer Lending___ Stock Type: ___Gazelle___

Profile: ___Offers credit cards, home equity loans and other consumer loans___

Address: ___201 Mission St., San Francisco, CA 94105___

CEO: ___Shailesh Mehta___ CFO: ___David Petrini___

Phone Number: ___(415) 543-0404___

Internet Address: ___www.providian___

Dividend Reinvestment Plan: ___Yes___ Contact: ___(800) 317-4445___

Direct Purchase Plan: ___Yes___ Contact: ___(800) 482-8690___

Average Annual Return '93-'98: ___49.8 %___

Name: __Qualcomm__          Symbol: __QCOM__     Exchange: __Nasdaq__

Industry: __Telecomm. Equipment__          Stock Type: __Gazelle__

Profile: ___Manufacturer of digital communications equipment___

Address: ___6455 Lusk Blvd., San Diego, CA 92121___

CEO: __Irwin Jacobs__          CFO: ___Anthony Thornley___

Phone Number: ___(619) 587-1121___     or     ___(619) 658-4224___

Internet Address: ___www.qualcomm.com___

Dividend Reinvestment Plan: __No__

Direct Purchase Plan: _____No_____

Average Annual Return '93-'98: ___14.9 %___

---

Name: ___Reliant___          Symbol: __REI__     Exchange: __NYSE__

Industry: __Electric Utility__          Stock Type: __Tortoise__

Profile: ___Operates electric utility in Texas___

Address: ___Box 4567, Houston, TX 77210___

CEO: __Don Jordan__          CFO: ___Stephen Naeve___

Phone Number: ___(713) 207-3000___

Internet Address: ___www.reliant.com___

Dividend Reinvestment Plan: __Yes__     Contact: ___(800) 231-6406___

Direct Purchase Plan: _____Yes_____     Contact: ___(800) 774-4117___

Average Annual Return '93-'98: ___13.5 %___

Name: ____Robert Half_____    Symbol: __RHI__    Exchange: __NYSE__

Industry: ___Business Services-Staffing_____    Stock Type: ___Gazelle____

Profile: ____Provider of personnel in finance, legal and information technology_____

Address: ___2884 Sand Hill Rd., Menlo Park, CA  94025_____

CEO: ___Harold Messmer, Jr._____    CFO: ___Keith Waddell_____

Phone Number: ___(650) 234-6000___

Internet Address: ___www.rhii.com_____

Dividend Reinvestment Plan: ___No_____

Direct Purchase Plan: _____No_____

Average Annual Return '93-'98: ____59.0 %____

---

Name: ___Sara Lee_____    Symbol: __SLE__    Exchange: __NYSE__

Industry: ___Food Products_____    Stock Type: ___Hare_____

Profile: ____Maker and marketer of branded foods_____

Address: ___Three First National Plaza, Chicago, IL  60602_____

CEO: ___John Bryan_____    CFO: ___Judith Sprieser_____

Phone Number: ___(312) 726-2600_____ or ___(800) 654-7272_____

Internet Address: ___www.saralee.com_____

Dividend Reinvestment Plan: ___Yes_____    Contact: ___(800) 554-3406_____

Direct Purchase Plan: _____No_____

Average Annual Return '93-'98: ____20.3 %____

Name: ___SBC Communications___ Symbol: ___SBC___ Exchange: ___NYSE___

Industry: ___Telecomm. Services___ Stock Type: ___Hare___

Profile: ___Provider of local and regional telecommunications services___

Address: ___Box 2933, San Antonio, TX  78299___

CEO: ___Edward Whitacre___ CFO: ___Donald Kiernan___

Phone Number: ___(210) 821-4105___

Internet Address: ___www.sbc.com___

Dividend Reinvestment Plan: ___Yes___ Contact: ___(800) 351-7221___

Direct Purchase Plan: ___Yes___ Contact: ___(800) 351-7221___

Average Annual Return '93-'98: ___24.8 %___

---

Name: ___Schering-Plough___ Symbol: ___SGP___ Exchange: ___NYSE___

Industry: ___Pharmaceuticals___ Stock Type: ___Gazelle___

Profile: ___Maker of prescription drugs and over the counter drugs___

Address: ___One Giralda Farms, Madison, NJ  07940___

CEO: ___Richard Kogan___ CFO: ___Jack Wyszomierski___

Phone Number: ___(973) 822-7000___

Internet Address: ___www.schering-plough.com___

Dividend Reinvestment Plan: ___Yes___ Contact: ___(800) 432-0140___

Direct Purchase Plan: ___No___

Average Annual Return '93-'98: ___48.3 %___

Name: __Schlumberger__    Symbol: __SLB__    Exchange: __NYSE__

Industry: __Oil Equipment & Services__    Stock Type: __Hare__

Profile: __Technological leader in oil services and drilling__

Address: __277 Park Ave., New York, NY 10172__

CEO: __Euan Baird__    CFO: __Jack Liu__

Phone Number: __(800) 997-5299__

Internet Address: __www.slb.com__

Dividend Reinvestment Plan: __No__

Direct Purchase Plan: __No__

Average Annual Return '93-'98: __11.3 %__

---

Name: __Schwab, Charles__    Symbol: __SCH__    Exchange: __NYSE__

Industry: __Securities Broker__    Stock Type: __Gazelle__

Profile: __Largest retail discount broker in U.S.__

Address: __101 Montgomery St., San Francisco, CA 94104__

CEO: __Charles Schwab__    CFO: __Steven Scheid__

Phone Number: __(415) 627-7000__ or __(800) 435-4000__

Internet Address: __www.schwab.com__

Dividend Reinvestment Plan: __Yes__    Contact: __(800) 468-9716__

Direct Purchase Plan: __No__

Average Annual Return '93-'98: __64.7 %__

Name: ___Solectron_____     Symbol: ___SLR___   Exchange: ___NYSE___

Industry: ___Electrical Manufacturer_____     Stock Type: ___Gazelle___

Profile: ___Provides contract electronic and assembly manufacture (outsource)___

Address: ___847 Gibraltar Dr., Milpitas, CA  95035_____

CEO: ___Koichi Nishimura_____     CFO: ___Susan Wang_____

Phone Number: ___(408) 956-6542___   or   ___(408) 957-8500___

Internet Address: ___www.solectron.com_____

Dividend Reinvestment Plan: ___No_____

Direct Purchase Plan: _____No_____

Average Annual Return '93-'98: ___45.6 %___

---

Name: ___Staples_____     Symbol: ___SPLS___   Exchange: ___Nasdaq___

Industry: ___Office Supplies_____     Stock Type: ___Gazelle___

Profile: ___Leading office supply retail chain of stores_____

Address: ___500 Staples Dr., Framingham, MA  01702_____

CEO: ___Thomas Stemberg_____     CFO: ___James Flavin_____

Phone Number: ___(508) 253-5000___   or   ___(800) 468-7751___

Internet Address: ___www.staples.com_____

Dividend Reinvestment Plan: ___No_____

Direct Purchase Plan: _____No_____

Average Annual Return '93-'98: ___54.0 %___

Name: ___Starbucks_____    Symbol: __SBUX__    Exchange: __Nasdaq__

Industry: ___Restaurants_____    Stock Type: ___Gazelle____

Profile: ____Leading roaster and retailer of coffee_____

Address: ___Box 34067, Seattle, WA 98124_____

CEO: ___Howard Schultz_____    CFO: ___Michael Casey_____

Phone Number: __(800) 239-0317___

Internet Address: ___www.starbucks.com_____

Dividend Reinvestment Plan: ___No_____

Direct Purchase Plan: _____No_____

Average Annual Return '93-'98: ___38.2 %____

---

Name: ___State Street_____    Symbol: ___STT___    Exchange: __NYSE__

Industry: ___Financial Services_____    Stock Type: ___Hare___

Profile: ____Provides accounting, custody, record keeping and Financial Services__

Address: ___225 Franklin St., Boston, MA 02101_____

CEO: ___Marshall Carter_____    CFO: ___Ronald O'Kelley_____

Phone Number: ___(617) 786-3000___ or ___(877) 639-7788_____

Internet Address: ___www.statestreet.com_____

Dividend Reinvestment Plan: ___Yes_____    Contact: ___(800) 426-5523___

Direct Purchase Plan: _____No_____

Average Annual Return '93-'98: ___31.9 %____

Name: ___Sun Microsystems___    Symbol: __SUNW__   Exchange: __Nasdaq__

Industry: ___Computers___                    Stock Type: ___Gazelle___

Profile: ___Supplier of workstations, servers, software and services___

Address: ___901 San Antonio Rd., Palo Alto, CA  94303___

CEO: ___Scott McNealy___          CFO: ___Michael Lehman___

Phone Number: ___(650) 960-1300___   or   ___(800) 801-7869___

Internet Address: ___www.sun.com___

Dividend Reinvestment Plan: ___No___

Direct Purchase Plan: _____No_____

Average Annual Return '93-'98: ___63.7 %___

Name: ___Tellabs___          Symbol: __TLAB__   Exchange: __Nasdaq__

Industry: ___Telecomm. Equipment___          Stock Type: ___Gazelle___

Profile: ___Maker of telecommunications equipment for digital networks___

Address: ___4951 Indiana Ave., Lisle, IL  60532___

CEO: ___Michael Birck___          CFO: ___Peter Galielmi___

Phone Number: ___(630) 378-8800___

Internet Address: ___www.tellabs.com___

Dividend Reinvestment Plan: ___No___

Direct Purchase Plan: _____No_____

Average Annual Return '93-'98: ___63.3 %___

Name: ___Tyco International___ Symbol: ___TYC___ Exchange: ___NYSE___

Industry: ___Manufacturing___ Stock Type: ___Hare___

Profile: ___Makes fire protection, flow control, medical and electronic devices___

Address: ___Headqtrs: Bermuda; U.S.: One Tyco Park, Exter, NH 03833___

CEO: ___Dennis Kozlowski___ CFO: ___Mark Swartz___

Phone Number: ___(603) 778-9700___

Internet Address: ___www.tycoint.com___

Dividend Reinvestment Plan: ___Yes___ Contact: ___(800) 685-4509___

Direct Purchase Plan: ___No___

Average Annual Return '93-'98: ___43.0 %___

---

Name: ___US Bancorp___ Symbol: ___USB___ Exchange: ___NYSE___

Industry: ___Banks___ Stock Type: ___Hare___

Profile: ___Regional bank operating in Midwest and Northwest U.S.___

Address: ___601 Second Avenue South, Minneapolis, MN 55402___

CEO: ___John Grandhofer___ CFO: ___Susan Lester___

Phone Number: ___(612) 973-2263___

Internet Address: ___www.usbank.com___

Dividend Reinvestment Plan: ___Yes___ Contact: ___(800) 446-2617___

Direct Purchase Plan: ___No___

Average Annual Return '93-'98: ___31.6 %___

Name: ___Vitesse_____     Symbol: _VTSS__   Exchange: _Nasdaq_

Industry: ___Semiconductor_____     Stock Type: ___Gazelle____

Profile: ___Makes digital high speed integrated circuits_____

Address: ___741 Calle Plano, Camatillo, CA  93012_____

CEO: ___Louis Tomasetta_____     CFO: ___Eugene Hovanec_____

Phone Number: ___(805) 388-3700____

Internet Address: ___www.vitesse.com_____

Dividend Reinvestment Plan: ___No_____

Direct Purchase Plan: _____No_____

Average Annual Return '93-'98: ___104.0 %____

---

Name: ___Walgreens_____     Symbol: _WAG__   Exchange: _NYSE__

Industry: ___Drug Stores_____     Stock Type: ___Hare_____

Profile: ___Largest U.S. drug store chain_____

Address: ___200 Wilmot Rd., Deerfield, IL  60015_____

CEO: ___L. Daniel Jorndt_____     CFO: ___Roger Polark_____

Phone Number: ___(847) 940-2500____

Internet Address: ___www.walgreens.com_____

Dividend Reinvestment Plan: ___Yes_____   Contact: ___(888) 368-7346____

Direct Purchase Plan: _____Yes_____   Contact: ___(888) 290-7264____

Average Annual Return '93-'98: ___43.6 %____

Name:    Walmart                Symbol:  WMT    Exchange:  NYSE

Industry:    Discount Retailer                Stock Type:    Hare

Profile:    World's largest retailer expanding internationally

Address:    Bentonville, AR  72716-8611

CEO:    David Glass                CFO:    John Menzer

Phone Number:    (501) 273-4000

Internet Address:    www.wal-mart.com

Dividend Reinvestment Plan:    Yes            Contact:    (800) 438-6278

Direct Purchase Plan:            Yes            Contact:    (800) 438-6278

Average Annual Return '93-'98:    27.6 %

---

Name:    Washington Mutual        Symbol:  WM    Exchange:  NYSE

Industry:    Savings & Loan                Stock Type:    Hare

Profile:    Largest thrift in the U.S.

Address:    1201 Third Ave., Seattle, WA  98101

CEO:    Kerry Killinger                CFO:    William Longbrake

Phone Number:    (206) 461-2000

Internet Address:    www.washingtonmutual.com

Dividend Reinvestment Plan:    Yes            Contact:    (800) 234-5835

Direct Purchase Plan:            No

Average Annual Return '93-'98:    22.3 %

Name: ___Waste Management___     Symbol: __WMI__     Exchange: __NYSE__

Industry: ___Environmental Services___          Stock Type: ___Hare___

Profile: ___Largest solid-waste disposal company in No. America___

Address: ___1001 Fannin St., Houston, TX 77002___

CEO: ___J.E. Drury___          CFO: ___E.E. DeFrates___

Phone Number: ___(713) 512-6000___

Internet Address: ___www.wastemanagement.com___

Dividend Reinvestment Plan: ___Yes___     Contact: ___(312) 461-5535___

Direct Purchase Plan: ___No___

Average Annual Return '93-'98: ___32.6 %___

---

Name: ___Weingarten Realty___     Symbol: __WRI__     Exchange: __NYSE__

Industry: __REIT__          Stock Type: ___Tortoise___

Profile: ___REIT holding shopping centers___

Address: ___2600 Citadel Plaza Dr., Houston, TX 77292___

CEO: ___Stanford Alexander___          CFO: ___J. Robertson___

Phone Number: ___(713) 866-6000___

Internet Address: ___www.weingarten.com___

Dividend Reinvestment Plan: ___Yes___     Contact: ___(800) 550-4689___

Direct Purchase Plan: ___Yes___     Contact: ___(888) 887-2966___

Average Annual Return '93-'98: ___10.3 %___

Name: __Xerox__    Symbol: __XRX__    Exchange: __NYSE__

Industry: ___Office Equipment___    Stock Type: __Hare__

Profile: __Makes, sells and services copiers, printers and office document machines__

Address: ___Box 1600, Stamford, CT 06904___

CEO: __Paul Allaire__    CFO: __Barry Romeril__

Phone Number: ___(203) 968-3000___    or    __(800) 828-6396__

Internet Address: ___www.xerox.com___

Dividend Reinvestment Plan: __Yes__    Contact: ___(800) 828-6396___

Direct Purchase Plan: _____No_____

Average Annual Return '93-'98: ___34.6 %___

---

Name: ___Xilinx___    Symbol: __XLNX__    Exchange: __Nasdaq__

Industry: ___Semiconductors___    Stock Type: __Hare__

Profile: ___Designs, develops and markets programmable logic integrated circuits___

Address: ___2100 Logic Dr., San Jose, CA 95124___

CEO: ___William Roelandts___    CFO: ___Gordon Steele___

Phone Number: ___(408) 559-7778___    or    __(800) 836-4002__

Internet Address: ___www.xilinx.com___

Dividend Reinvestment Plan: __No__

Direct Purchase Plan: _____No_____

Average Annual Return '93-'98: ___32.5 %___

# Bibliography

Barber, Brad and John Lyon. *Firm Size, Book to Market Ratio, Price Momentum, and Earnings Momentum.* Davis: University of California, 1997.

Barber, Brad and Terrance Odean. "Trading is Hazardous to Your Wealth." *Journal of Finance* (1999).

Bernstein, Peter L. *Against the Gods.* New York: Wiley, 1996.

Bernstein, Peter L. and Aswath Damodaran. *Investment Management.* New York: Wiley, 1998.

Breally, Richard, Stewart Meyers, and Alan Marcus. *Fundamentals of Corporate Finance.* New York: McGraw-Hill, 1999.

Collins, James and Jerry Porras. *Built to Last.* New York: HarperCollins, 1997.

de Geus. *The Living Company.* Boston: Harvard Business School Press, 1997.

Dembo, Ron S. and A. Freeman. *Seeing Tomorrow.* New York: Wiley, 1998.

Dreman, David. *Contrarian Investment Strategies.* New York: Simon & Schuster, 1998.

Fama, Eugene and Kenneth French. "Common Risk Factors in the Returns on Stocks and Bonds." *Journal of Financial Economics.* 1993: pp. 3–56.

Gallea, Anthony and William Patalon. *Contrarian Investing.* New York: New York Institute of Finance, 1996.

Hagstrom, Robert. *The Warren Buffett Way.* New York: Wiley, 1994.

Hagstrom, Robert. *The Warren Buffett Portfolio.* New York: Wiley, 1999.

Hamel, Gary and C. K. Prahalad. *Competing for the Future.* Boston: Harvard Business School Press, 1994.

Kalb, Scott and Peggy Kalb. *The Top 100 International Growth Stocks.* New York: Fireside, 1998.

Malkiel, Burton. A *Random Walk Down Wall Street.* New York: Norton, 1996.

Malkiel, Burton. *Global Bargain Hunting.* New York: Simon & Schuster, 1998.

Moore, Geoffrey, Paul Johnson, and Tom Kippola. *The Gorilla Game.* New York: Harper Business, 1998.

O'Shaughnessy, James. *What Works on Wall Street.* New York: McGraw-Hill, 1997.

Rappaport, Alfred. *Creating Shareholder Value.* New York: The Free Press, 1998.

Rosenweig, Jeffrey. *Winning the Global Game.* New York: The Free Press, 1998.

Sharpe, William, Gordon Alexander, and Jeffrey Bailey. *Investments,* 5th ed. Saddle River, NJ: Prentice-Hall, 1995.

Siegel, Jeremy. *Stocks for the Long Run.* New York: McGraw-Hill Irwin, 1994.

Slywotzky, Adrian and Davis Morrison. *The Profit Zone.* New York: Times Business, 1997.

Stewart, Thomas A. *Intellectual Capital.* New York: Doubleday, 1997.

Stewart, G. Bennett. *The Quest for Value.* New York: HarperCollins, 1991.

Teece, David J. "Capturing Value from Knowledge Assets." *California Management Review,* (Spring 1998): 55–76.

Tigue, Joseph and Joseph Lisanti. *The Dividend Rich Investor.* New York: McGraw-Hill, 1999.

Tully, Shawn. *Greatest Wealth Creators. Fortune* (November 9, 1998): 193–204.

Value Line, *Value Line Selection.* (Febuary 12, 1999): 5734.

# Appendix A

## Glossary of Terms

**arbitrage:** The simultaneous purchase and sale of substantially identical assets in order to profit from a price difference in two markets.

**asset:** Something of monetary value that is owned by a firm or an individual.

**balance sheet:** Record of a company's finances at a given time.

**bandwidth:** A measure of the information capacity in bits/second or MegaHz.

**barrier to entry:** An obstacle to entry to an industry or market.

**beta:** A measure of a stock's volatility expressed as how much its price moves in relation to the performance of the stock market.

**blue chip:** The tried and true performers in the stock market.

**brand:** The attribution to a product of a name, trademark, or symbol.

**brand equity:** The collection of all assets linked to a brand.

**business cycle:** A fluctuation in the level of economic activity which forms an expansion followed by a contraction.

**capacity:** The amount that a plant or enterprise can produce in a given period.

**capital:** Financial assets held by the firm.

**cash flow:** Amount of net cash flowing into a company during a specified period. The sum of retained earnings and depreciation provision.

**competitive advantage:** Strengths that a company holds as an advantage over its competitors.

**conglomerate:** A holding company that owns companies in a wide range of different businesses or industries.

**contrarian:** An investor who is willing to think and act differently from the commonly accepted viewpoint.

**cost of capital:** The weighted average of the cost of a firm's debt and equity capital.

**core competence:** The particular activity or know-how that an enterprise is significantly good at.

**debt:** An obligation arising from borrowing or purchasing on credit.

**demand:** The amount that people are ready to purchase at the prevailing price.

**depreciation:** The reduction in value of an asset due to wear and use.

**economies of scale:** The decrease in the unit cost of production as a firm increases its volume of production.

**economies of scope:** The decrease in the unit cost of production as a firm increases its range of related products.

**efficient market:** A market in which prices reflect all available information and adjust rapidly to any new information.

**entrepreneur:** A person who seeks to create wealth and lasting value by starting and developing a new venture.

**equity:** An ownership interest in a company, often held as shares in the capital of the company.

**financial strength rating:** A measure of the relative financial strength of a company as rated by Value Line.

**fixed asset:** An asset that remains in the business over time such as buildings and land.

**fixed cost:** A cost that does not vary with the quantity of goods produced over the short run (near term).

**generally accepted accounting principles (GAAP):** The official financial reporting criteria for U. S. firms.

**global business:** Activities that take advantage of worldwide opportunities and resources.

**gross margin:** The difference between revenues and the cost of goods sold.

**income statement:** A summary of a firm's revenues and experiences for a specified period.

**increasing returns:** See law of increasing returns.

**inflation:** A sustained rise in the price of goods and services.

**initial public offering:** The initial sale of common stock to individuals in the public market.

**innovation:** The act of introducing a new process or product into a market.

**intangible asset:** A nonphysical asset such as patents, copyrights, and business secrets.

**intellectual capital:** Knowledge that has been formalized, captured, and used to produce a process that provides a significant value-added product or service.

**intrinsic value:** The true or real value of a common stock.

**inventory:** Finished goods held for use or sale by a business firm.

**law of diminishing returns:** The principle that as increasing incremental resources are put into an activity, the less are the resulting incremental increases in outputs.

**law of increasing returns:** The principle that for certain industries, such as software, that as increasing incremental resources are put into an activity, the incremental increases in output will grow.

**learning organization:** A firm that uses its corporate experiences to improve and adapt.

**leverage:** The use of borrowed funds to increase the return on an investment. A firm's debt-to-equity ratio is a measure of its leverage.

**liquidity:** Assets that are cash or can be easily converted to money.

**market capitalization:** The current market price of one share of a company's common stock multiplied by the number of shares outstanding.

**market timing:** The timely shifting of assets into or out of the market.

**net present value (NPV):** The current value of future amounts of money, discounted at an agreed rate (discount rate).

**net worth:** Total assets minus total liabilities. Also known as shareholder equity.

**operating margin:** Earnings before interest, taxes, depreciation, and amortization divided by sales and expressed as a percentage.

**option:** A contract allowing another party to buy or sell a given item within a given time period at an agreed price.

**outsourcing:** A firm's purchase of services or goods from suppliers rather than doing them within the firm.

**price growth persistence index:** A measure of the historic tendency of a stock to show persistent growth compared to the average stock.

**price stability index:** A measure of the stability of a stock price relative to the market changes.

**productivity:** The ratio of the output obtained to the inputs utilized for a given process.

**price/cash flow:** Price divided by cash flow per share. Cash flow equals net income plus noncash expenses such as depreciation and deferred taxes.

**real estate investment trust:** A firm that invests its capital in income producing real estate.

**return on equity:** Net income divided by equity expressed as a percentage.

**risk:** The potential for a loss of revenues or profits due to the occurrence of an event. The potential for a loss of an investment.

**risk premium:** The higher return that an investment should provide for its higher risk.

**safety rank:** A measure of the potential risk (safety) of a stock provided by Value Line.

**technological progress:** The involvement of economic growth which enables more output to be produced for unchanged inputs of labor and capital. This improvement is attributed to the development and use of new technologies.

**timing:** See market timing

**total return:** The total percentage a stock has returned, including price appreciation and dividend payouts.

**utility:** A public service company that provides electric, gas, or water to its customers.

**value added:** The value of a process output minus the value of its inputs.

**value chain:** The sequence of steps or subprocesses that a firm uses to produce its product or service.

# Appendix B

## Internet Addresses for Investors

| Name | Address |
|---|---|
| Aluminum Industry | www.aluminum.org |
| American Association of Individual Investors | www.aaii.org |
| | |
| Biotechnology Industry | www.bio.org |
| *Business Week* Magazine | www.businessweek.com |
| | |
| Cents Financial | lp-llc.com/cents |
| Citizens Social Investing | www.citizensfunds.com |
| CBS Market Watch—News | cbs.marketwatch.com |
| Charts | www.bigcharts.com |
| CNBC | www.cnbc.com |
| Computers | www.idc.com |
| | |
| Direct Purchase of Stocks | www.enrolldirect.com |
| Dismal Scientist—Economics | www.dismalscientist.com |
| Dividend Reinvestment Plans-List of Plans and Links | www.dripcentral.com |
| | |
| Earnings Estimates | www.zacks.com |
| Economic Value Added—EVA | www.eva.com |
| *Economist* Magazine | www.economist.com |
| Efficient Market Theory | www.efficientfrontier.com |
| | |
| Hoover Online | www.hoovers.com |
| | |
| Ibbotson Associates—Data Source | www.ibbotson.com |
| Individual Investor Magazine | www.iionline.com |
| Internet Stocks | www.internetstocks.com |
| Investment Clubs | www.better-investing.org |
| | |
| Market Guide—Stock Data | www.marketguide.com |
| Morningstar Stocks | www.morningstar.net |

| Name | Address |
| --- | --- |
| *New York Times* | www.nytimes.com |
| Netstock Direct—List of Direct Purchase Plans and Dividend Reinvestment Plans | www.netstock-direct.com |
| Oil and Gas Industry | www.oillink.com |
| Quicken—Stock data search | www.quicken.com |
| Real Estate Investment Trusts | www.nareit.com |
| S&P Personal Wealth | www.personalwealth.com |
| *San Jose Mercury* Newspaper | www.mercurycenter.com |
| Semiconductor Industry | www.sia.org |
| Social Investing | www.socialinvest.org |
| Software Stocks | www.softwarestocks.com |
| Standard & Poor | www.stockinfo.standardpoor.com |
| Technology | www.techstocks.com |
| Upside Magazine | www.upside.com |
| *Wired* Magazine—Internet Stocks | Stocks.wired.com |
| Yahoo Finance—Stock Data | quote.yahoo.com www.yahoo.com |
| Yardeni, Ed, Economist | www.yardeni.com |

# Index